CITES Orchid Checklist

Volume 3

For the genera:

Aerangis, *Angraecum*, *Ascocentrum*, *Bletilla*, *Brassavola*, *Calanthe*, *Catasetum*, *Miltonia*, *Miltonioides, Miltoniopsis, Renanthera, Renantherella, Rhynchostylis, Rossioglossum, Vanda* and *Vandopsis*

Compiled by:

Jacqueline A Roberts, Sharon Anuku, Joanne Burdon, Paul Mathew, H Noel McGough & Andrew Newman

Assisted by a selected international panel of orchid experts

Royal Botanic Gardens, Kew

First published in 2001

General editor of series: Jacqueline A Roberts

ISBN 1 84246 033 1

Produced with the financial assistance of
the CITES Nomenclature Committee and
the Royal Botanic Gardens, Kew

Cover design by Media Resources RBG Kew
Printed in Great Britain by The Cromwell Press

Acknowledgements / Remerciements / Reconocimientos

The compilers would like to thank their colleagues at the Royal Botanic Gardens, Kew and the following orchid experts for their help with the preparation of the checklist for publication. All suggestions for amendments were gratefully noted and included at the discretion of the compilers. We would particularly like to thank **Dr Mark Clements, Mr Paul Ormerod** and **Dr Wesley Higgins** for their comprehensive review of the final version of the text. Finally, we are very grateful to the CITES Secretariat for translating the text into French and Spanish.

Les compilateurs tiennent à remercier leurs collègues des *Royal Botanic Gardens* de Kew et les experts en orchidées suivants, pour leur aide dans la préparation de la Liste en vue de sa publication. Il a été pris note avec gratitude de toutes les suggestions, qui ont été incluses à la discrétion des compilateurs. Nours adressons des remerciements particuliers à **Dr Mark Clements, M. Paul Ormerod** et **Dr Wesley Higgins**, qui ont examiné de manière approfondie la version finale du texte. Nous remercions vivement le Secrétariat CITES d'avoir traduit le texte en espagnol et en français.

Las personas encargadas de la recopilación de este volumen desean dar las gracias a sus colegas del *Royal Botanic Gardens,* Kew, así como a los siguientes especialistas en orquídeas por su ayuda en la preparación de la presente lista. Se tomó nota con agradecimiento de todas las sugerencias de enmienda, que fueron incluidas conforme a los criterios de los responsables de este volumen. En particular deseamos dar las gracias al **Dr Mark Clements, Sr. Paul Ormerod** y **Dr Wesley Higgins** por la revisión de la versión definitiva del texto. Agradecemos a la Secretaría CITES por la traducción del texto al español y al francés.

International Panel of Orchid Experts
Groupe international de spécialistes des orchidées
Grupo Internacional de Expertos en Orquídeas

Leonid Averyanov	Russia
Germán Carnevali	Mexico
James Comber	UK
Phillip Cribb	UK
Isobyl la Croix	UK
Richard Evenden	UK
Olaf Gruss	Germany
Johan Hermans	UK
Rudolf Jenny	Switzerland
Gavin McDonald	South Africa
David Menzies	UK
Gustavo Romero	USA
Chen Sing-chi	China
Manfred Wolff	Germany
Tsi Zhang-Huo	China
Mark Clements	Australia
Paul Ormerod	Australia
Wesley Higgins	USA

CONTENTS

TABLE DES MATIERES

INDICE

Preámbulo

CITES CHECKLIST - ORCHIDACEAE

PREAMBLE

1. Background

The 1992 Conference of the Parties to the Convention on International Trade in Endangered Species of Wild Fauna and Flora (CITES) adopted Resolution Conf. 8.19 which called for the production of a standard reference to the names of Orchidaceae.

The Vice-Chairman of the CITES Nomenclature Committee was charged with the responsibility of co-ordinating the input needed to produce such a reference.

The orchid genera identified as priorities in the *Review of Significant Trade in Species of Plants included in Appendix II of CITES* (CITES Doc. 8.31) would be treated first. The checklists (or parts thereof) as they came available would be put to the Conference of the Parties for approval.

At its third meeting (Chiang Mai, Thailand, November 1992) the Plants Committee extensively discussed a proposal by the Vice-Chairman of the Nomenclature Committee regarding the possible mechanisms to develop the Standard Reference. The Plants Committee endorsed a procedure by which compilations from the available literature, made on a central database, would be circulated to a panel of international experts for consultation and final decisions on the valid names to be used for the taxa concerned.

During the development of the checklists every effort was made to recruit national experts from the range states using the contact network of the regional representatives of the Plants Committee.

Based on the recommendations of the Plants Committee and on those of Resolution Conf. 8.19, the CITES Secretariat established a Memorandum of Understanding with the Royal Botanic Gardens, Kew for the preparation of this reference. The work started on 1 July 1993. Recommendation 6 of the *Review of Significant Trade of Species of Plants included in Appendix II of CITES* outlined the following genera as priorities:

Aerangis, Angraecum, Ascocentrum, Bletilla, Brassavola, Calanthe, Catasetum Cattleya, Coelogyne, Comparettia, Cymbidium, Cypripedium, Dendrobium, Disa, Dracula, Encyclia, Epidendrum, Laelia, Lycaste, Masdevallia, Miltonia, Miltoniopsis, Odontoglossum, Oncidium, Paphiopedilum, Paraphalaenopsis, Phalaenopsis, Phragmipedium, Renanthera, Rhynchostylis, Rossioglossum, Sophronitis, Vanda and *Vandopsis.*

The Memorandum of Understanding stated that completed and verified checklists of the following groups should be prepared for consideration by the ninth meeting of the Conference of the Parties:

Cattleya, Cypripedium, Laelia, Paphiopedilum, Phalaenopsis, Phragmipedium, Pleione and *Sophronitis.*

This work was completed in 1995, resulting in the publication of *CITES Orchid Checklist Volume 1*, which also included accounts of the genera *Constantia, Paraphalaenopsis* and *Sophronitella.*

In February 1996 a second Memorandum of Understanding stated that completed and verified checklists should be prepared for the following groups:

Cymbidium, Dendrobium (selected sections only), *Disa, Dracula* and *Encyclia.*

This work was completed in 1997, resulting in the publication of *CITES Orchid Checklist Volume 2.*

Volumes 1 and 2 are the adopted CITES standards to be used as a guideline when making reference to the species names of the genera concerned.

Volume 3 was initiated in September 1998 with a Letter of Agreement between the CITES Secretariat and the Royal Botanic Gardens, Kew, stating that completed and verified checklists should be prepared for the following groups:

Angraecum, Aerangis, Ascocentrum, Bletilla, Brassavola, Calanthe, Catasetum, Miltonia, Renanthera, Rhynchostylis, Rossioglossum, Vanda and *Vandopsis*

The Vice-Chairman of the Nomenclature Committee reported on progress at the ninth (Darwin, Australia, June 1999) and tenth (Shepherdstown, West Virginia, December 2000) meetings of the Plants Committee. The Eleventh Meeting of the Conference of the Parties to CITES approved the continuing process and adopted this volume as the guideline when making reference to the species names of the genera concerned.

This list is only a guideline, is subject to change and is not to be taken or treated as the definitive word on the taxonomy and systematics of these plants.

2. Computer aspects

Hardware: The database was set up on a Networked Pentium II PC using ALICE software, Version 2.1.

Database system: The ALICE database system was used to handle the data collection. ALICE handles distribution, uses, common names, descriptions, habitats, synonymy, bibliography and other classes of data for species, subspecies or varieties. It allows users to design their own reports for checklists, studies, monographs or conservation lists, for example.

Alice Software can be contacted by e-mail at info@alicesoftware.com
Web page: http://www.alicesoftware.com

3. Compilation procedures

- Primary references were identified by orchid specialists based at the Royal Botanic Gardens, Kew.
- An International Panel of Orchid Experts was established in order to review each stage of the checklist.
- Information was entered into the ALICE taxonomic database and a preliminary report produced.
- Preliminary reports for each genus were distributed to the Panel of Orchid Experts for their comments on any additions or amendments needed.
- Additions and amendments returned from the Panel members were entered into the database. These were linked to a reference contained in the bibliography at the end of the report on each genus. (Not included in this checklist, but copies held for reference at RBG, Kew).
- This sequence was repeated five times for each genus to allow full consultation with the Panel.

- A 'final draft' including all the genera was prepared and distributed to further experts for their comments.
- Dr Mark Clements, Dr Wesley Higgins and Mr. Paul Ormerod were consulted as external editors to review the final draft.
- Any further additions and amendments were added to the database.
- Format for publication was agreed with the CITES Secretariat and reports generated using AWRITE and prepared for camera-ready copy using Microsoft Word for Windows version 97.

4. Conservation

During the consultation process with the Panel of Orchid Experts, information was also requested on the conservation status of the species concerned. Copies of the present and proposed new IUCN categories of threat were distributed to the Panel for their use.

5. How to use the checklist

It is intended that this Checklist be used as a quick reference for checking accepted names, synonymy and distribution. The reference is therefore divided into three main parts:

Part I: All names in current use

An alphabetical list of all accepted names and synonyms included in this checklist - a total of 2279 names (817 accepted and 1462 synonyms).

Part II: Accepted names in current use

Separate lists for each genus. Each list is ordered alphabetically by the accepted name and details are given on current synonyms and distribution.

Part III: Country checklist

Accepted names from all genera included in this checklist are ordered alphabetically under country of distribution.

6. Conventions employed in parts I, II and III

a) Accepted names are presented in **bold roman** type.
Synonyms are presented in *italic* type.

b) Duplicate names

In Part I, the author's name appears after each taxon where the taxon name appears twice or more e.g., *Amblyglottis veratrifolia* Blume, *Amblyglottis veratrifolia* (Willd.) Blume, (unless the author's name is the same).

i) Where a synonym occurs more than once, but refers to different species, for example, *Amblyglottis veratrifolia* (a synonym of both **Calanthe ceciliae** and **Calanthe triplicata**), the name with an asterisk is the species most likely to be encountered in trade. For example:

All Names	Accepted Name
Amblyglottis speciosa	**Calanthe speciosa**
Amblyglottis veratrifolia	**Calanthe ceciliae**
Amblyglottis veratrifolia	**Calanthe triplicata***
Angorchis alcicornis	**Aerangis alcicornis**

*Species most likely to be in trade (in this example, **Calanthe triplicata**).

ii) Where an accepted name and a synonym are the same, but refer to different species, for example, **Angraecum viride** (accepted name) and *Angraecum viride* (a synonym of **Angraecum rhynchoglossum**), the name with an asterisk is the species most likely to be seen in trade. For example:

All Names	Accepted Name
Angraecum virgula	**Tridactyle virgula**
Angraecum viride	
Angraecum viride	**Angraecum rhynchoglossum***
Angraecum viridescens	**Tridactyle laurentii**

*Species most likely to be in trade (in this example, **Angraecum rhynchoglossum**).

NB: In examples bi) and bii) it is necessary to double-check by reference to the distribution as detailed in Part II. For instance, in the example bii), if the name given was 'Angraecum viride' and it was known that the plant in question came from Madagascar, this would indicate that the species was **Angraecum rhynchoglossum**, being traded under the synonym *Angraecum viride*. **Angraecum viride** is only found in Kenya and Tanzania.

c) Natural hybrids have been included in the checklist and are indicated by the multiplication sign ×. They are arranged alphabetically within the lists.

7. Number of names entered for each genus:
Aerangis (Accepted: 56, Synonyms: 162); *Angraecum* (Accepted: 237, Synonyms: 229); *Ascocentrum* (Accepted: 16, Synonyms: 19); *Bletilla* (Accepted: 6, Synonyms: 42); *Brassavola* (Accepted: 16, Synonyms: 55); *Calanthe* (Accepted: 203, Synonyms: 302); *Catasetum* (Accepted: 158, Synonyms: 324); *Miltonia* (Accepted: 14, Synonyms: 38); *Miltonioides* (Accepted: 7, Synonyms: 27); *Miltoniopsis* (Accepted: 7, Synonyms: 26); *Renanthera* (Accepted: 16, Synonyms: 23); *Renantherella* (Accepted: 1, Synonyms: 2); *Rhynchostylis* (Accepted: 4, Synonyms: 58); *Rossioglossum* (Accepted: 11, Synonyms: 34); *Vanda* (Accepted: 59, Synonyms: 98) and *Vandopsis* (Accepted: 6, Synonyms: 23).

8. Geographical areas
Country names follow the United Nations standard as laid down in *Country Names. Terminology Bulletin* No. 347/Rev. 1, 1997. United Nations.

9. Orchidaceae controlled by CITES
The family Orchidaceae is listed on Appendix II of CITES. In addition the following taxa are listed on Appendix I at time of publication:

Cattleya trianaei
Dendrobium cruentum
Laelia jongheana
Laelia lobata
Paphiopedilum spp.
Peristeria elata
Phragmipedium spp.
Renanthera imschootiana
Vanda coerulea

10. Abbreviations, botanical terms, and Latin*

Not all these abbreviations, botanical terms and Latin will appear in this Checklist; however, they have been included as a useful reference.

Note: words in *italics* are Latin

ambiguous name a name which has been applied to different taxa by different authors, so that it has become a source of ambiguity
anon. anonymous; without author or author unknown
auct. *auctorum*: of authors
CITES Convention on International Trade in Endangered Species of Wild Fauna and Flora
cultivar an individual, or assemblage of plants maintaining the same distinguishing features, which has been produced or is maintained (propagated) in cultivation
cultivation the raising of plants by horticulture or gardening; not immediately taken from the wild
descr. *descriptio*: the description of a species or other taxonomic unit
distribution where plants are found (geographical)
ed. editor
edn. edition (book or journal)
eds. editors
epithet the last word of a species, subspecies, or variety (etc.), for example: *aurantiacum* is the species epithet for the species *Ascocentrum aurantiacum* and *philippinense* is the subspecific epithet for *Ascocentrum aurantiacum* ssp. *philippinense*
escape a plant that has left the boundaries of cultivation (e.g. a garden) and is found occurring in natural vegetation
ex *ex*: after; may be used between the name of two authors, the second of whom validly published the name indicated or suggested by the first
excl. *exclusus*: excluded
forma *forma*: a taxonomic unit inferior to variety
hort. *hortorum*: of gardens (horticulture); raised or found in gardens; not a plant of the wild
ICBN International Code for Botanical Nomenclature
in prep. in preparation
in sched. *in scheda*: on a herbarium specimen or label
in syn. *in synonymia*: in synonymy
incl. including
ined. *ineditus*: unpublished
introduction a plant which occurs in a country, or any other locality, due to human influence (by purpose or chance); any plant which is not native
key a written system used for the identification of organisms (e.g. plants)
leg. *legit*: he gathered; the collector
misspelling a name that has been incorrectly spelt; not a new or different name
morphology the form and structure of an organism (e.g. a plant)
name causing confusion a name that is not used because it cannot be assigned unambiguously to a particular taxon (e.g. a species of plant)
native an organism (e.g. a plant) that occurs naturally in a country, or region, etc.
naturalized a plant which has either been introduced (see introduction) or has escaped (see escape) but which looks like a wild plant and is capable of reproduction in its new environment
nom. ambig. *nomen ambiguum*: ambiguous name

nom. cons. prop. *nomen conservandum propositum*: name proposed for conservation under the rules of the International Code for Botanical Nomenclature (ICBN)
nom. illeg *nomen illegitimum*: illegitimate name
nom. *nomen*: name
nom. nud. *nomen nudum*: name published without description
nomenclature branch of science concerned with the naming of organisms (e.g. plants)
non *non*: not
only known from cultivation a plant which does not occur in the wild, only in cultivation
orthographic variant an alternative spelling for the same name
p.p. *pro parte*: partly, in part
pro parte *pro parte*: partly, in part
provisional name name given in anticipation of a valid description
sens. lat. *sensu lato*: in the broad sense; a taxon (usually a species) and all its subordinate taxa (e.g. subspecies) and/or other taxa sometimes considered as distinct
sens. *sensu*: in the sense of; the manner in which an author interpreted or used a name
sensu *sensu*: in the sense of; the manner in which an author interpreted or used a name
sic *sic*, used after a word that looks wrong or absurd, to show that it has been quoted correctly
spp. species
ssp. subspecies
synonym a name that is applied to a taxon but which cannot be used because it is not the accepted name – the synonym or synonyms form the synonymy
taxa plural of taxon
taxon a named unit of classification, e.g. genus, species, subspecies
var. variety

*thanks to Dr Aaron Davis, RBG Kew, for the provision of this guide

11. Bibliography

Primary reference sources used in the compilation of checklists:

Ball, J.S. (1978). *Southern African Epiphytic Orchids*. Johannesburg: Conservation Press. pp248.

Bechtel, H., Cribb, P.J. & Launert, G.O.E (1992). *The Manual of Cultivated Orchid Species*. 3rd ed. London: Blandford. pp 585.

Cadet, J. (1989). *Joyaux de nos Forets: Les Orchidees de la Reunion*. Saint Denis: Ile de la Reunion: Nouvelle Impr. Dionysienne

Carnevali, G. & Ramírez, I.M. (1999). *Brassavola* In Carnevali, G., Ramírez, I.M., Romero, G. & Vargas, C. (1999). *Orchidaceae* in Steyermark, J.A., Berry, P.E. & Holst, B.K. (eds.) *Flora of the Venezuelan Guyana*. 5. St Louis: Missouri Botanical Garden. pp833.

Lang Kaiyong, Chen Singchi, Luo Yibo & Zhu Guanghua (1999). *Flora Reipublicae Popularis Sinicae. Tomus 17: Angiospermae: Monocotyledoneae: Orchidaceae (1)*. Beijing: Science Press. pp463.

Chen Singchi, Tsi Zhanhuo, Lang Kaiyong & Zhu Guanghua (1999). *Flora Reipublicae Popularis Sinicae. Tomus 18: Angiospermae: Monocotyledoneae: Orchidaceae* (*2*). Beijing: Science Press. pp551.

Chen Singchi, Tsi Zhanhuo, & Zhu Guanghua (1999). *Flora Reipublicae Popularis Sinicae. Tomus 19: Angiospermae: Monocotyledoneae: Orchidaceae* (*3*). Beijing: Science Press. pp485.

Clements, M.A. (1982). *Preliminary Checklist of Australian Orchidaceae*. Canberra, Australia: National Botanic Gardens. pp216.

Colombian Orchid Society (1990). *Native Colombian Orchids: Volume 1 Acacallis - Dryadella*. Medellin, Colombia: Compania Litografica National S.A.

Colombian Orchid Society (1991). Native Colombian Orchids: Volume 3 Maxillaria - Ponthieva. Medellin, Colombia: Compania Litografica National S.A.

Comber, J.B. (1990). *Orchids of Java*. Bentham-Moxon Trust, Royal Botanic Gardens, Kew, UK. pp407.

Cribb, P.J. & Robbins, S. (1991). The Genus Bletilla in Cultivation. *Orchid Review 99*, 406-409.

Cribb, P.J. (1984). *Flora of Tropical East Africa: Orchidaceae 3*. Balkema, Rotterdam.

Dockrill, A.W. (1992). *Australian Indigenous Orchids Volume 1: The Epiphytes, the tropical terrestrial species*. Chipping Norton, NSW: Surrey Beatty and Sons. pp524.

Dressler, R.L. (1980). *Orchids of Panama*. Missouri Botanical Garden, USA.

Dunsterville, G.C.K. & Garay, L.A. (1976). *Venezuelan Orchids Illustrated*. 6. A.Deutsch, London.

Dunsterville, G.C.K. (1987). *Venezuelan Orchids*. Armitano, Venezuela.

DuPuy, D. et al. (1999). *The Orchids of Madagascar: annotated checklist*. Royal Botanic Gardens, Kew, UK. pp376.

Grove, D.L. (1995). *Vandas & Ascocendas & their combinations with other genera*. Portland, Oregon: Timber Press. pp241.

Karasawa, K. (1998). *Orchid Atlas Volume 5: Calanthe - Coelogyne*. Orchid Atlas Publishing Society, Tokyo.

La Croix, I. & E. (1997). *African Orchids in the Wild and Cultivation*. Portland, Oregon: Timber Press. pp379.

La Croix, I. et al. (1991). *Orchids of Malawi: the epiphytic and terrestrial orchids from South and East Central Africa*. Rotterdam: A.A. Balkema. pp358.

Pabst, G.F.J. & Dungs, F. (1975). *Orchidaceae Brasilienses*. Ed. 1. Hildesheim: Brucke-Verlag Schmersow.

Perrier de la Bathie, H. (1981). *Flora of Madagascar (Vascular Plants): 49th family Orchids*. Lodi, California: S.D. Beckman. pp542.

Richard, A. H. (1974). *Flora of the Lesser Antilles: Leeward and Windward Islands*. Jamaica Plain, Mass.: Arnold Arboretum, Harvard University.

Seidenfaden, G. & Wood, J.J. (1992). *The Orchids of Peninsular Malaysia and Singapore - A Revision of R.E. Holttum: Orchds of Malaya*. Fredensborg, Denmark: Olsen & Olsen.

Seidenfaden, G. (1975). Orchid Genera in Thailand I - III. *Dansk Botanisk Arkiv* 33(3). 1-228.

Seidenfaden, G. (1992). The Orchids of Indochina. *Opera Botanica* 114, Copenhagen.

Seidenfaden, G. (1988). Orchid Genera in Thailand XIV. Fifty-nine vandoid Genera. *Opera Botanica* 95, Copenhagen.

Senghas, K. (1997). *Miltonia und verwandte Gattungen*. Schweizerische Orchideen Gesellschaft, Zurich. pp119.

Stewart, J. & Campbell, B. (1996). *Orchids of Kenya*. Winchester: St Paul's Bibliographies,. pp176.

Stewart, J. (1979). Revision of the African species of Aerangis (Orchidacea). *Kew Bulletin* 34: 2.

Wiard, L.A. (1987). *An Introduction to the Orchids of Mexico*. Ithaca, N.Y: Comstock Pub. Assocs,. pp239.

Williams, L.O. (1951). The Orchidaceae of Mexico. *CEIBA*, 2.

Williams, L.O. (1956). An enumeration of the Orchidaceae of Central America, British Honduras & Panama. *CEIBA*, 5.

Wood, J.J. & Cribb, P.J. (1994). *A Checklist of the Orchids of Borneo*. Royal Botanic Gardens, Kew, UK. pp409.

LISTE CITES DES ORCHIDACEAE

PREAMBULE

1. Contexte

En 1992, la Conférence des Parties à la Convention sur le commerce international des espèces de faune et de flore sauvages menacées d'extinction (CITES) a adopté la résolution Conf. 8.19 dans laquelle elle recommande la préparation d'une liste normalisée de référence des noms d'Orchidaceae.

Le vice-président du Comité CITES de la nomenclature a été chargé de coordonner les informations reçues en vue de préparer cette liste.

Les genres d'orchidées classés comme prioritaires dans l'Etude du commerce important d'espèces végétales inscrites à l'Annexe II de la CITES (document CITES Doc. 8.31) devaient être les premiers traités. Les listes (ou parties de listes) devaient être soumises à la Conférence des Parties pour approbation à mesure qu'elles seraient disponibles.

A sa troisième session (Chiang Mai, Thaïlande, novembre 1992) le Comité pour les plantes a abondamment discuté d'une proposition du vice-président du Comité de la nomenclature concernant les mécanismes possibles d'élaboration d'une liste de référence normalisée. Le Comité pour les plantes a approuvé une procédure par laquelle les compilations faites à partir de la littérature disponible, sur une base de données centrale, seraient envoyées à un groupe d'experts internationaux pour consultation et décision finale sur les noms valides devant être utilisés pour les taxons concernés.

Au cours de la préparation des listes, il a été fait appel aux spécialistes nationaux des Etats de l'aire de répartition, en utilisant le réseau de contacts des représentants régionaux du Comité pour les plantes. Toutefois, il y a eu peu de réponses. L'on espère que la publication de la présente Liste favorisera une participation accrue aux futurs volumes.

Sur la base des recommandations du Comité pour les plantes et de la résolution Conf. 8.19, le Secrétariat CITES a établi un protocole d'accord avec les Jardins botaniques royaux de Kew pour la préparation de la liste de référence. Le travail a débuté le 1er juillet 1993. La recommandation 6 de l'Etude du commerce important d'espèces végétales inscrites à l'Annexe II de la CITES recommande de considérer les genres suivants comme prioritaires:

Aerangis, Angraecum, Ascocentrum, Bletilla, Brassavola, Calanthe, Catasetum Cattleya, Coelogyne, Comparettia, Cymbidium, Cypripedium, Dendrobium, Disa, Dracula, Encyclia, Epidendrum, Laelia, Lycaste, Masdevallia, Miltonia, Miltoniopsis, Odontoglossum, Oncidium, Paphiopedilum, Paraphalaenopsis, Phalaenopsis, Phragmipedium, Renanthera, Rhynchostylis, Rossioglossum, Sophronitis, Vanda et *Vandopsis.*

Le protocole d'accord prévoyait que des listes seraient établies et vérifiées pour les groupes suivants, et soumises pour examen à la neuvième session de la Conférence des Parties:

Cattleya, Cypripedium, Laelia, Paphiopedilum, Phalaenopsis, Phragmipedium, Pleione et *Sophronitis.*

Préambule

Cette tâche a été achevée en 1995 et a abouti à la publication de la *CITES Orchid Checklist Volume 1*, qui inclut également des données sur les genres *Constantia, Paraphalaenopsis* et *Sophronitella.*

En février 1996, un second protocole d'accord a porté sur l'établissement et la vérification de listes pour les groupes suivants:

Cymbidium, Dendrobium (certaines sections sélectionnées), *Disa, Dracula* et *Encyclia.*

Cette tâche a été achevée en 1997 et a abouti à la publication de la *CITES Orchid Checklist Volume 2.*

Les Volumes 1 et 2 constituent la liste de référence CITES adoptée, à utiliser lorsqu'on se réfère aux noms des espèces de ces genres.

La préparation du Volume 3 a commencé en septembre 1998 suite à un accord passé entre le Secrétariat CITES et les Jardins botaniques royaux de Kew selon lequel des listes complétées et vérifiées devraient être préparées pour les groupes suivants:

Angraecum, Aerangis, Ascocentrum, Bletilla, Brassavola, Calanthe, Catasetum, Miltonia, Renanthera, Rhynchostylis, Rossioglossum, Vanda and *Vandopsis*

Le vice-président du Comité de la nomenclature a présenté un rapport d'activité aux neuvième (Darwin, Australia, juin 1999) et dixième (Shepherdstown, West Virginia, décembre 2000) sessions du Comité pour les plantes. La 11e session de la Conférence des Parties à la CITES a approuvé la poursuite du processus et a adopté le présent volume comme ligne directrice pour l'utilisation des noms d'espèces des genres concernés.

Cette liste fait office de ligne directrice; elle peut être modifiée et ne doit pas être comprise ou utilisée comme faisant autorité pour la taxonomie et la systématique de ces plantes.

2. Aspects informatiques

Matériel: La base de données a été établie en réseau sur un PC Pentium II en utilisant le logiciel ALICE, version 2.1.

Système de base de données: Le système ALICE a été utilisé pour enregistrer les données. ALICE traite la répartition géographique, les utilisations, les noms communs, les descriptions, les habitats, les synonymes, la bibliographie et d'autres catégories de données relatives aux espèces, sous-espèces ou variétés. Il permet aux utilisateurs de créer leurs propres rapports sous forme de listes, d'études, de monographies, de listes de conservation, etc.

Alice Software peut être contacté par courriel à: info@alicesoftware.com
Sur Internet: http://www.alicesoftware.com

3. Procédure de compilation

- Les principales références ont été identifiées par les spécialistes des orchidées des Jardins botaniques royaux de Kew.
- Un groupe international de spécialistes des orchidées a été établi afin d'examiner la Liste à chaque étape.

- Des informations ont été entrées dans la base de données taxonomiques ALICE et un rapport préliminaire a été préparé.
- Des rapports préliminaires sur chaque genre ont été remis au groupe de spécialistes pour qu'il formule ses commentaires sur les additions ou amendements nécessaires.
- Les ajouts et amendements des membres du groupe ont été entrés dans la base de données. Ils ont été reliés à une référence incluse dans la bibliographie à la fin du rapport sur chaque genre. (Non incluse dans la présente Liste mais dont des copies sont à disposition, pour référence, aux JBR à Kew).
- Le processus a été répété cinq fois pour chaque genre afin de permettre la pleine consultation du groupe.
- Un "projet final" incluant tous les genres a été préparé et envoyé à d'autres spécialistes en leur demandant leurs commentaires.
- MM. Mark Clements, Wesley Higgins et Paul Ormerod ont été consultés en tant que réviseurs externes et ont examiné le projet final.
- Tous les autres ajouts et amendements ont été ajoutés à la base de données.
- La présentation retenue pour la publication a été convenue avec le Secrétariat CITES et les rapports ont été créés en utilisant AWRITE et préparés pour la publication en utilisant Microsoft Word pour Windows, version 97.

4. Conservation

Au cours du processus de consultation du groupe de spécialistes, des informations ont été demandées sur l'état de conservation des espèces concernées. Des copies des catégories actuelles et des nouvelles catégories de menaces proposées par l'UICN ont été distribuées au groupe.

5. Comment utiliser la Liste?

Cette Liste devrait être utilisée comme liste de référence pour vérifier rapidement les noms acceptés, les synonymes et la répartition géographique. Elle est divisée en trois parties principales:

Première partie: Tous les noms d'usage courant
Liste alphabétique de tous les noms acceptés et des synonymes inclus dans la Liste, 2279 noms (817 noms acceptés et 1462 synonymes).

Deuxième partie: Noms acceptés d'usage courant
Listes séparées pour chaque genre. Chaque liste est donnée dans l'ordre alphabétique des noms acceptés et comporte des indications sur les synonymes et la répartition géographique actuels.

Troisième partie: Liste des pays
Les noms acceptés de tous les genres inclus dans cette liste sont donnés par ordre alphabétique sous chaque pays de l'aire de répartition.

6. Conventions utilisées dans les première, deuxième et troisième parties

a) Les noms acceptés sont imprimés **bold roman**.
Les synonymes sont en *italique*.

b) Noms identiques pour des taxons différents:

Dans la première partie, le nom de l'auteur apparaît après chaque taxon lorsque le taxon apparaît au moins deux fois; exemple: *Amblyglottis veratrifolia* Blume, *Amblyglottis veratrifolia* (Willd.) Blume, (sauf si le nom de l'auteur est le même).

i) Lorsque le synonyme apparaît au moins une fois mais renvoie à des noms acceptés différents – par exemple, *Amblyglottis veratrifolia* (synonyme à la fois de **Calanthe ceciliae** et de **Calanthe triplicata**), le nom comportant un astérisque est celui de l'espèce la plus susceptible d'être trouvée dans le commerce. Exemple:

Tous les noms	Noms acceptés
Amblyglottis speciosa	**Calanthe speciosa**
Amblyglottis veratrifolia	**Calanthe ceciliae**
Amblyglottis veratrifolia	**Calanthe triplicata***
Angorchis alcicornis	**Aerangis alcicornis**

*Espèce la plus susceptible d'être trouvée dans le commerce (dans cet exemple, **Calanthe triplicata**)

ii) Lorsque le nom accepté et un synonyme sont les mêmes mais renvoient à des espèces différentes – par exemple **Angraecum viride** (nom accepté) et *Angraecum viride* (synonyme de **Angraecum rhynchoglossum**), le nom comportant un astérisque est celui de l'espèce la plus susceptible d'être trouvée dans le commerce. Exemple:

Tous les noms	Noms acceptés
Angraecum virgula	**Tridactyle virgula**
Angraecum viride	
Angraecum viride	**Angraecum rhynchoglossum***
Angraecum viridescens	**Tridactyle laurentii**

*Espèce la plus susceptible d'être trouvée dans le commerce (dans cet exemple, **Angraecum rhynchoglossum**)

NB: Dans les exemples bi) et bii), il faut faire une double vérification en se référant à la répartition géographique indiquée dans la deuxième partie. Ainsi, dans l'exemple bii), si le nom donné est 'Angraecum viride' et si l'on sait que la plante vient de Madagascar, cela indique qu'il s'agit de **Angraecum rhynchoglossum**, commercialisée sous le synonyme *Angraecum viride.* **Angraecum viride** ne pousse qu'au Kenya et en République-Unie de Tanzanie.

c) Les hybrides naturels figurent dans les listes, dans l'ordre alphabétique, et sont indiqués par le signe de multiplication ×.

7. Nombre de noms entrés pour chaque genre:

Aerangis (acceptés: 56, synonymes: 162); *Angraecum* (acceptés: 237, synonymes: 229); *Ascocentrum* (acceptés: 16, synonymes: 19); *Bletilla* (acceptés: 6, synonymes: 42); *Brassavola* (acceptés: 16, synonymes: 55); *Calanthe* (acceptés: 203, synonymes: 302); *Catasetum* (acceptés: 158, synonymes: 324); *Miltonia* (acceptés: 14, synonymes: 38); *Miltonioides* (acceptés: 7, synonymes: 27); *Miltoniopsis* (acceptés: 7, synonymes: 26);

Renanthera (acceptés: 16, synonymes: 23); *Renantherella* (acceptés: 1, synonymes: 2); *Rhynchostylis* (acceptés: 4, synonymes: 58); *Rossioglossum* (acceptés: 11, synonymes: 34); *Vanda* (acceptés: 59, synonymes: 98) and *Vandopsis* (acceptés: 6, synonymes: 23).

8. Régions géographiques
Les noms des pays sont ceux figurant dans le *Country Names. Terminology Bulletin* No. 347/Rev. 1, 1997. United Nations.

9. Orchidées soumises aux contrôles CITES
La famille des Orchidaceae est inscrite à l'Annexe II de la CITES. De plus, les taxons suivants étaient inscrits à l'Annexe I au moment de la publication de la Liste:

Cattleya trianaei
Dendrobium cruentum
Laelia jongheana
Laelia lobata
Paphiopedilum spp.
Peristeria elata
Phragmipedium spp.
Renanthera imschootiana
Vanda coerulea

10. Abréviations, termes botaniques, et mots en latin *
Ces termes de botanique, noms latins et abréviations ne sont pas tous utilisés dans la Liste. Ils sont inclus pour référence.

Note: les mots *en italique* sont d'origine latine

ambiguous name (nom ambigu) nom donné à différents taxons par différents auteurs, ce qui crée une ambiguïté
anon. anonyme; sans auteur
auct. *auctorum*: d'auteurs
CITES Convention sur le commerce international des espèces de faune et de flore sauvages menacées d'extinction
cultivar spécimen ou groupe de plantes conservant les mêmes caractéristiques distinctives, produites ou conservées (propagées) en culture
cultivation (culture) obtention de plantes par horticulture ou jardinage, par opposition au prélèvement dans la nature
descr. *descriptio* description d'une espèce ou d'une autre entité taxonomique
distribution (aire de répartition géographique) région(s) où se trouve les plantes
ed. éditeur
edn. édition (d'un livre ou d'un périodique)
eds. éditeurs
epithet (épithète) dernier mot d'une espèce, d'une sous-espèce ou d'une variété (etc.). Exemple: *aurantiacum* est l'épithète de l'espèce *Ascocentrum aurantiacum* et *philippinense* l'épithète infraspécifique de *Ascocentrum aurantiacum* ssp. *philippinense*
escape (échappée) qualifie une plante qui a quitté l'enceinte de culture (jardin, par exemple) et qu'on retrouve dans la végétation naturelle
ex *ex* d'après; peut être utilisé entre deux noms d'auteurs, dont le second a validement publié le nom d'après les indications ou suggestions du premier
excl. *exclusus* exclu

hort. *hortorum* de jardins (horticole); plante cultivée ou se trouvant dans des jardins horticoles, par opposition à une plante d'origine sauvage
ICBN (CINB) Code international de la nomenclature botanique
in prep. en préparation
in sched. *in scheda* sur un spécimen d'herbier ou une étiquette
in syn. *in synonymia* en synonymie
incl. incluant
ined. *ineditus* non publié
introduction résultat d'une activité humaine (volontaire ou non) aboutissant à ce qu'une plante non indigène se retrouve dans un pays ou une région
key (clé) système écrit utilisé pour la détermination d'organismes (plantes, par exemple)
leg. *legit* il ramassa; le collecteur
misspelling (faute d'orthographe) nom mal orthographié, par opposition à un nom nouveau ou différent
morphology (morphologie) forme et structure d'un organisme (d'une plante, par exemple)
name causing confusion (nom causant une confusion) nom qui n'est pas utilisé parce qu'il ne peut être assigné sans ambiguïté à un taxon particulier (à une espèce de plante, par exemple)
native (indigène) qualifie un organisme (une plante, par exemple) prospérant naturellement dans un pays ou une région etc.
naturalized (naturalisée) qualifie une plante introduite (voir introduction) ou échappée (voir échappée) qui ressemble à une plante sauvage et qui se propage dans son nouvel environnement
nom. *nomen* nom
nom. ambig. *nomen ambiguum* nom ambigu
nom. cons. prop. *nomen conservandum propositum* nom dont le maintien a été proposé d'après les règles du *International Code of Botanical Nomenclature* (Code international de la nomenclature botanique)
nomenclature branche de la science qui nomme les organismes (les plantes, par exemple)
non *non* pas
only known from cultivation (connue seulement en culture) qualifie une plante qu'on ne trouve pas à l'état sauvage
orthographic variant (variante orthographique) même nom orthographié différemment
pro parte *pro parte* partiellement, en partie
provisional name (nom provisoire) nom donné par anticipation d'une description
sens. *sensu* au sens de; manière dont un auteur interprète ou utilise un nom
sens. lat. *sensu lato* au sens large; un taxon (habituellement une espèce) et tous ses taxons inférieurs (sous-espèce, etc.) et/ou d'autres taxons parfois considérés comme distincts
sic *sic*, utilisé après un mot qui semble faux ou absurde; indique que ce mot est cité textuellement
synonym (synonyme) nom donné à un taxon mais qui ne peut être utilisé parce que ce n'est pas le nom accepté; le ou les synonymes forment la synonymie
taxa pluriel de taxon
taxon unité taxonomique à laquelle on a attribué un nom - genre, espèce, sous-espèce, etc.
var. variété

* Nous remercions M. Aaron Davis, des JBR de Kew, d'avoir fourni ce guide

11. Bibliographie

Principales sources de références utilisées dans la compilation des listes:

Ball, J.S. (1978). *Southern African Epiphytic Orchids.* Johannesburg: Conservation Press. pp248.

Bechtel, H., Cribb, P.J. & Launert, G.O.E (1992). *The Manual of Cultivated Orchid Species.* 3rd ed. London: Blandford. pp 585.

Cadet, J. (1989). *Joyaux de nos Forets: Les Orchidees de la Reunion.* Saint Denis: Ile de la Reunion: Nouvelle Impr. Dionysienne

Carnevali, G. & Ramírez, I.M. (1999). *Brassavola* In Carnevali, G., Ramírez, I.M., Romero, G. & Vargas, C. (1999). *Orchidaceae* in Steyermark, J.A., Berry, P.E. & Holst, B.K. (eds.) *Flora of the Venezuelan Guyana.* 5. St Louis: Missouri Botanical Garden. pp833.

Lang Kaiyong, Chen Singchi, Luo Yibo & Zhu Guanghua (1999). *Flora Reipublicae Popularis Sinicae. Tomus 17: Angiospermae: Monocotyledoneae: Orchidaceae (1).* Beijing: Science Press. pp463.

Chen Singchi, Tsi Zhanhuo, Lang Kaiyong & Zhu Guanghua (1999). *Flora Reipublicae Popularis Sinicae. Tomus 18: Angiospermae: Monocotyledoneae: Orchidaceae (2).* Beijing: Science Press. pp551.

Chen Singchi, Tsi Zhanhuo, & Zhu Guanghua (1999). *Flora Reipublicae Popularis Sinicae. Tomus 19: Angiospermae: Monocotyledoneae: Orchidaceae (3).* Beijing: Science Press. pp485.

Clements, M.A. (1982). *Preliminary Checklist of Australian Orchidaceae.* Canberra, Australia: National Botanic Gardens. pp216.

Colombian Orchid Society (1990). *Native Colombian Orchids: Volume 1 Acacallis - Dryadella.* Medellin, Colombia: Compania Litografica National S.A.

Colombian Orchid Society (1991). Native Colombian Orchids: Volume 3 Maxillaria - Ponthieva. Medellin, Colombia: Compania Litografica National S.A.

Comber, J.B. (1990). *Orchids of Java.* Bentham-Moxon Trust, Royal Botanic Gardens, Kew, UK. pp407.

Cribb, P.J. & Robbins, S. (1991). The Genus Bletilla in Cultivation. *Orchid Review 99*, 406-409.

Cribb, P.J. (1984). *Flora of Tropical East Africa: Orchidaceae 3.* Balkema, Rotterdam.

Dockrill, A.W. (1992). *Australian Indigenous Orchids Volume 1: The Epiphytes, the tropical terrestrial species.* Chipping Norton, NSW: Surrey Beatty and Sons. pp524.

Dressler, R.L. (1980). *Orchids of Panama*. Missouri Botanical Garden, USA.

Dunsterville, G.C.K. & Garay, L.A. (1976). *Venezuelan Orchids Illustrated*. 6. A.Deutsch, London.

Dunsterville, G.C.K. (1987). *Venezuelan Orchids*. Armitano, Venezuela.

DuPuy, D. et al. (1999). *The Orchids of Madagascar: annotated checklist*. Royal Botanic Gardens, Kew, UK. pp376.

Grove, D.L. (1995). *Vandas & Ascocendas & their combinations with other genera*. Portland, Oregon: Timber Press. pp241.

Karasawa, K. (1998). *Orchid Atlas Volume 5: Calanthe - Coelogyne*. Orchid Atlas Publishing Society, Tokyo.

La Croix, I. & E. (1997). *African Orchids in the Wild and Cultivation*. Portland, Oregon: Timber Press. pp379.

La Croix, I. et al. (1991). *Orchids of Malawi: the epiphytic and terrestrial orchids from South and East Central Africa*. Rotterdam: A.A. Balkema. pp358.

Pabst, G.F.J. & Dungs, F. (1975). *Orchidaceae Brasilienses*. Ed. 1. Hildesheim: Brucke-Verlag Schmersow.

Perrier de la Bathie, H. (1981). *Flora of Madagascar (Vascular Plants): 49th family Orchids*. Lodi, California: S.D. Beckman. pp542.

Richard, A. H. (1974). *Flora of the Lesser Antilles: Leeward and Windward Islands*. Jamaica Plain, Mass.: Arnold Arboretum, Harvard University.

Seidenfaden, G. & Wood, J.J. (1992). *The Orchids of Peninsular Malaysia and Singapore - A Revision of R.E. Holttum: Orchds of Malaya*. Fredensborg, Denmark: Olsen & Olsen.

Seidenfaden, G. (1975). Orchid Genera in Thailand I - III. *Dansk Botanisk Arkiv* 33(3). 1-228.

Seidenfaden, G. (1992). The Orchids of Indochina. *Opera Botanica* 114, Copenhagen.

Seidenfaden, G. (1988). Orchid Genera in Thailand XIV. Fifty-nine vandoid Genera. *Opera Botanica* 95, Copenhagen.

Senghas, K. (1997). *Miltonia und verwandte Gattungen*. Schweizerische Orchideen Gesellschaft, Zurich. pp119.

Stewart, J. & Campbell, B. (1996). *Orchids of Kenya*. Winchester: St Paul's Bibliographies,. pp176.

Stewart, J. (1979). Revision of the African species of Aerangis (Orchidacea). *Kew Bulletin* 34: 2.

Wiard, L.A. (1987). *An Introduction to the Orchids of Mexico.* Ithaca, N.Y: Comstock Pub. Assocs,. pp239.

Williams, L.O. (1951). The Orchidaceae of Mexico. *CEIBA*, 2.

Williams, L.O. (1956). An enumeration of the Orchidaceae of Central America, British Honduras & Panama. *CEIBA*, 5.

Wood, J.J. & Cribb, P.J. (1994). *A Checklist of the Orchids of Borneo.* Royal Botanic Gardens, Kew, UK. pp409.

LISTA CITES - ORCHIDACEAE

PREÁMBULO

1. Antecedentes

En 1992 la Conferencia de las Partes en la Convención sobre el Comercio Internacional de Especies Amenazadas de Fauna y Flora Silvestres (CITES) aprobó la Resolución Conf. 8.19, en la que se solicita que se prepare una referencia normalizada sobre los nombres de las Orchidaceae.

Se encargó al Vicepresidente del Comité de Nomenclatura de la CITES que coordinase las tareas necesarias para preparar dicha referencia.

Se acordó abordar en primer lugar los géneros de orquídeas identificados como prioritarios en el Examen del Comercio Signifitativo de especies incluidas en el Apéndice II de la CITES (CITES Doc. 8.31). Las listas (o partes de las mismas) se presentarían a la aprobación de la Conferencia de las Partes a medida que se fuesen preparando.

En su tercera reunión (Chiang Mai, Tailandia, noviembre de 1992), el Comité de Flora examinó detenidamente una propuesta del Vicepresidente del Comité de Nomenclatura sobre los posibles mecanismos para preparar un referencia normalizada. El Comité de Flora ratificó un procedimiento mediante el cual se efectuarían recopilaciones a partir de las publicaciones disponibles sobre el tema, que se introducirían en una base central de datos, y se presentaría a un Grupo Internacional de Expertos para efectuar consultas y tomar decisiones sobre los nombres que deberían utilizarse para los taxa en cuestión.

Durante la preparación de las listas se desplegaron esfuerzos para contactar a expertos nacionales de los Estados del área de distribución, utilizando como coordinadores a los representantes regionales ante el Comité de Flora.

En virtud de las recomendaciones del Comité de Flora y las que figuran en la Resolución Conf. 8.19, la Secretaría de la CITES estableció un Memorándum de Entendimiento con el Real Jardín Botánico de Kew, para preparar dicha referencia. Las tareas se iniciaron el 1 de julio de 1993. En la Recomendación 6 del *Examen del Comercio Significativo de especies de plantas incluidas en el Apéndice II de la CITES*, se hace hincapié en dar prioridad a los siguientes géneros:

Aerangis, Angraecum, Ascocentrum, Bletilla, Brassavola, Calanthe, Catasetum Cattleya, Coelogyne, Comparettia, Cymbidium, Cypripedium, Disa, Dendrobium, Dracula, Encyclia, Epidendrum, Laelia, Lycaste, Masdevallia, Miltonia, Miltoniopsis, Odontoglossum, Oncidium, Paphiopedilum, Paraphalaenopsis, Phalaenopsis, Phragmipedium, Renanthera, Rhynchostylis, Rossioglossum, Sophronitis, Vanda y *Vandopsis.*

En el Memorándum de Entendimiento se indicaba que deberían prepararse listas completas y verificadas para los grupos siguientes para someterlas a la consideración de la novena reunión de la Conferencia de las Partes:

Cattleya, Cypripedium, Laelia, Paphiopedilum, Phalaenopsis, Phragmipedium, Pleione y *Sophronitis.*

Preámbulo

Esta labor se completó en 1995, dando como resultado la publicación del *CITES Orchid Checklist Volume 1*, en el que se incluyen también los géneros *Constantia, Paraphalaenopsis* y *Sophronitella.*

En febrero de 1996 se estableció un segundo Memorándum de Entendimiento en el que se dejaba constancia de que se prepararían listas completas y verificadas para los grupos siguientes:

Cymbidium, Dendrobium (únicamente las secciones seleccionadas), *Disa, Dracula* y *Encyclia.*

Esta labor se completó en 1997, dando como resultado la publicación del *CITES Orchid Checklist Volume 2.*

Los Volúmenes 1 y 2 son las referencias CITES adoptadas para su utilización como directrices al hacer referencia a los nombres de las especies de los géneros en cuestión.

El Volumen 3 se inició en septiembre de 1998 mediante una Carta de Acuerdo entre la Secretaría CITES y el Real Jardín Botánico de Kew, en la que se estipulaba que se prepararían Listas completas y verificadas para los siguientes grupos:

Angraecum, Aerangis, Ascocentrum, Bletilla, Brassavola, Calanthe, Catasetum, Miltonia, Renanthera, Rhynchostylis, Rossioglossum, Vanda and *Vandopsis.*

El Vicepresidente del Comité de Nomenclatura informó sobre los progresos realizados en las reuniones novena y décima del Comité de Flora (Darwin, Australia, junio de 1999) y (Shepherdstown, West Virginia, diciembre de 2000). En la 11a. reunión de la Conferencia de las Partes se aprobó que continuara este proceso y se adoptó el presente volumen como directriz al hacer referencia a los nombres de las especies de los generos de que se trata.

Esta lista es exclusivamente una directriz, está sujeta a modificaciones y no debe considerarse o tratarse como si fuese la obra definitiva sobre la taxonomía y la sistemática de estas plantas.

2. Programa informático

Soporte físico: La base de datos se instaló en un Compaq LTE Lite 4/25C computadora portátil equipada con un software ALICE.

Sistema de base de datos: Se utilizó el sistema de base de datos ALICE para la recopilación de los datos. ALICE se ocupa de la distribución, utilización, nombres comunes, descripciones, hábitat, sinónimos, bibliografía y otras clases de datos para las especies, subespecies o variedades. Permite a los usuarios preparar sus propios informes para las listas, estudios, monografías o listas de conservación.

El soporte lógico Alice puede contactarse por correo electrónico a la dirección info@alicesoftware.com
Página en la WEB:http://www.alicesofware. com

3. Procedimiento para la recopilación

- Las referencias preliminares fueron identificadas por especialistas en orquídeas del Real Jardín Botánico de Kew.
- Se estableció un Grupo Internacional de Expertos sobre orquídeas para que revisara cada una de las fases de la lista.

- Se introdujo la información en la base de datos taxonómica de ALICE y se preparó un informe preliminar.
- Los informes preliminares respecto de cada género se distribuyeron al Grupo de expertos sobre orquídeas para que formulase comentarios sobre cualquier adición o enmienda.
- Las adiciones y enmiendas remitidas por el Grupo de expertos se introdujeron en la base de datos, vinculándolas a una referencia contenida en la bibliografía al final del informe sobre cada género. (No incluida en la presente lista de control, pero se guardan copias para referencia en el Real Jardín Botánico de Kew).
- Esta secuencia se repitió cinco veces para cada género, a fin de realizar consultas pormenorizadas con el Grupo.
- Se preparó un "borrador final" en el que se incluían todos los géneros y se distribuyó a otros expertos para que formulasen comentarios.
- El Dr. Mark Clements, Dr. Wesley Higgins y el Mr. Paul Ormerod participaron como editores externos para revisar la versión final.
- Las nuevas adiciones o enmiendas se incluyeron en la base de datos.
- El formato para la publicación se acordó con la Secretaría de la CITES, los informes se efectuaron en AWRITE y el material preparado para la cámara se preparó utilizando Microsoft Word for Windows, versión 6.

4. Conservación

Durante el proceso de consultas con el Grupo de expertos sobre orquídeas, se solicitó también información sobre el estado de conservación de las especies en cuestión. Se transmitió al Grupo copia de las categorías de amenaza de la UICN actuales y propuestas.

5. Cómo emplear esta lista

La idea es que esta lista se utilice como referencia rápida para controlar los nombres aceptados, los sinónimos y la distribución. Así, pues, la referencia se divide en tres partes principales:

Parte I: Todos los nombres utilizados normalmente

Una lista por orden alfabético de todos los nombres y sinónimos aceptados - un total de 2279 nombres (817 aceptados y 1462 sinónimos).

Parte II: Nombres aceptados utilizados normalmente

Listas separadas para cada género. En cada lista se presentan por orden alfabético los nombres aceptados, con información sobre los sinónimos y la distribución.

Parte III: Lista por paises

Los nombres aceptados para todos los géneros incluidos en esta lista se presentan por orden alfabético según el país de distribución.

6. Sistema de presentación utilizado en las Partes I, II y III

a) Los nombres aceptados se presentan en tipo de letra negrita y romano. Los sinónimos se presentan en letra cursiva.

b) Nombres duplicados

En la Parte I, el nombre del autor aparece después de cada taxón, cuando dicho taxón se cita en más de una ocasión p.e. *Amblyglottis veratrifolia* Blume, *Amblyglottis veratrifolia* (Willd.) Blume, (a menos de que el nombre del autor sea el mismo).

i) Cuando un sinónimo aparece más de una vez, pero se refiere a diferentes nombres aceptados, a saber, *Amblyglottis veratrifolia*, (un sinónimo de ambas **Calanthe ceciliae** y **Calanthe triplicata)**, el nombre acompañado de un asterisco se refiere a la especie que con mayor probabilidad se encontrará en el comercio. Por ejemplo:

Todos los nombres	**Nombre aceptado**
Amblyglottis speciosa	**Calanthe speciosa**
Amblyglottis veratrifolia	**Calanthe ceciliae**
Amblyglottis veratrifolia	**Calanthe triplicata***
Angorchis alcicornis	**Aerangis alcicornis**

*La especie que con mayor probabilidad se encontrará en el comercio (en este ejemplo, **Calanthe triplicata**).

ii) Cuando un nombre aceptado es igual al sinónimo, pero se refiere a especies diferentes, a saber, **Angraecum viride** (nombre aceptado) and *Angraecum viride* (un sinónimo de **Angraecum rhynchoglossum**), el nombre acompañado de un asterisco se refiere a la especie que con mayor probabilidad se encontrará en el comercio. Por ejemplo:

Todos los nombres	**Nombre aceptado**
Angraecum virgula	**Tridactyle virgula**
Angraecum viride	
Angraecum viride	**Angraecum rhynchoglossum***
Angraecum viridescens	**Tridactyle laurentii**

*La especie que con mayor probabilidad se encontrará en el comercio (en este ejemplo, **Angraecum rhynchoglossum**).

NB: En los ejemplos b) i) y b) ii) es preciso efectuar doble verificación en lo que concierne a la distribución, como se indica en la Parte II. Por ejemplo, en el caso c), si el nombre dado fue "Angraecum viride" y se sabe que la planta en cuestión procede de Madagascar, querrá decir que la especie era **Angrecum rhynchoglossum**, que se comercializa bajo el sinónimo *Angrecum viride*, ya que **Angraecum viride** no se encuentra únicamente en Kenya y Tanzanía.

d) En la lista se han incluido los híbridos naturales y se indican con el signo de multiplicar "×". Se presentan por orden alfabético.

7. Número de nombres incluidos para cada género:
Aerangis (Aceptados: 56, Sinónimos: 162); *Angraecum* (Aceptados: 237, Sinónimos: 229); *Ascocentrum* (Aceptados: 16, Sinónimos: 19); *Bletilla* (Aceptados: 6, Sinónimos: 42); *Brassavola* (Aceptados: 16, Sinónimos: 55); *Calanthe* (Aceptados: 203, Sinónimos: 302); *Catasetum* (Aceptados: 158, Sinónimos: 324); *Miltonia* (Aceptados: 14, Sinónimos: 38); *Miltonioides* (Aceptados: 7, Sinónimos: 27); *Miltoniopsis* (Aceptados: 7, Sinónimos: 26); *Renanthera* (Aceptados: 16, Sinónimos: 23); *Renantherella* (Aceptados: 1, Sinónimos: 2); *Rhynchostylis* (Aceptedos: 4, Sinónimos: 58); *Rossioglossum* (Aceptados: 11, Sinónimos: 34); *Vanda* (Aceptados: 59, Sinónimos: 98) and *Vandopsis* (Aceptados: 6, Sinónimos: 23).

8. Áreas geográficas
Para los nombres de los países se ha seguido la referencia oficial de las Naciones Unidas. *Country Names. Terminology Bulletin* No. 347/Rev. 1, 1997. United Nations.

9. Orchidaceae controladas por la CITES
La familia de Orchidaceae está incluida en el Apéndice II de la CITES. Además, en el momento de esta publicación, están incluidos en el Apéndice I los siguientes taxa:

Cattleya trianaei
Dendrobium cruentum
Laelia jongheana
Laelia lobata
Paphiopedilum spp.
Peristeria elata
Phragmipedium spp.
Renanthera imschootiana
Vanda coerulea

10. Abreviaciones, términos botánicos, y expresiones latinas *

En esta Lista no aparecen todas las abreviaturas, términos botánicos y en latín, pese a que se han incluido como referencia útil.

Nota: las expresiones latinas aparecen en *bastardilla*

ambiguous name (nombre ambiguo) un nombre utilizado por distintos autores para diferentes taxa, de manera que da motivo a confusión
anon. Anonymous; autor desconocido
auct. *auctorum* de autores
CITES Convención sobre el Comercio Internacionale de Especies Amenazadas de Fauna y Flora Silvestres
cultivation (cultivo) el cultivo de plantas mediante hortícultura o jardinería; no se ha recolectado inmediantemente del medio silvestre
cultivar un ejemplar, o una agrupación de plantas, que tienen los mismos rasgos característicos, que ha sido producido o se mantiene (reproduce) en cultivo
descr. *descriptio* la descripción de una especie o de otra unidad taxonómica
distribution (distribución) donde se encuentran las plantas (geográfica)
ed. editor
edn. edición (libro o revista)
eds. editores
eptithet (epíteto) la última palabra de una especie, subespecie o variedad (etc.), por ejemplo: *aurantiacum* es el epíteto de la especie *Ascocentrum aurantiacum* y *philippinense* es el epíteto subespecífico de *Ascocentrum aurantiacum* ssp. *philippinense.*
escape (volverse silvestre) una planta que ha sobrepasado los límites del cultivo (p.e.: un jardín) y prospera en la naturaleza
ex *ex* después, puede utilizarse entre los nombres de dos autores, el segundo de los cuales publicó el nombre indicado o sugerido por el primero
excl. *exclusus* excluida
hort. *hortorum* de jardines (horticultura); cultivadas o prosperan en jardines; no se trata de una planta silvestre
ICNB (CINB) Código Internacional de Nomenclatura Botánica

incl. inclusive
in prep. en preparación
in sched. *in scheda* en un espécimen de herbario o etiqueta
in syn. *in synonymia* en sinonimia
ined. *ineditus* : inédito
introduction (introducción) una planta que ocurre en un país, o en cualquier otra localidad, debido a la influencia antropogénica (intencionalmente o al azar); cualquier planta que no es nativa
key (clave) un sistema escrito utilizado para la identificación de organismos (p.e.: plantas)
leg. *legit* el recolector; el coleccionista
misspelling (error de ortografía) un nombre que se ha escrito incorrectamente; no se trata de un nombre nuevo o diferente
morphology (morfología) la forma y estructura de un organismo (p.e.: una planta)
nombre que crea confusión: un nombre que no se usa, ya que su utilización crearía confusión
name causing confusion (nombre de dudosa semejanza) un nombre que no se usa, ya que no puede asignarse a un determinado taxón sin crear confusión (p.e.: una especie de planta)
native (nativo) un organismo (p.e.: una planta) que prospera naturalmente en un país o región, etc.
naturalized (naturalizada) una planta que ha sido introducida (véase introducción) o se ha vuelto silvestre (véase volverse silvestre) pero que parece una planta silvestre y se reproduce por sí misma en su nuevo medio
nom. *nomen* nombre
nom. ambig. *nomen ambiguum* nombre ambiguo
nom. cons. prop. *nomen conservandum propositum* nombre propuesto para la conservación con arreglo a lo dispuesto en el Código Internacional de Nomenclatura Botánica (ICBN)
nomenclature (nomenclatura) parte de la ciencia que se ocupa de atribuir nombres a organismos (p.e.: plantas)
non *non* no
only known from cultivation (solo se conoce en cultivo) una planta que no ocurre en el medio silvestre, únicamente en cultivo
orthographic variant (variante ortográfica) una alternativa ortográfica del mismo nombre
pro parte *pro parte* : parcialmente, en parte
provisional name (nombre provisional) nombre asignado temporalmente hasta que se disponga de una descripción válida
sens. *sensu* en el sentido de; la forma en que un autor interpreta o utiliza un nombre
sens. lat. *sensu lato* en sentido generalizado, un taxón (normalmente una especie) y todos sus taxa subordinados (p.e.: subespecies) y/o otros taxa a veces considerados como distintos
sic *sic* utilizado después de una palabra que pudiera parecer inexacta o absurda, para dar a entender que es textual
synonym (sinónimo) un nombre que se aplica a un taxón pero que no puede utilizarse ya que no es un nombre aceptado – el sinónimo o los sinónimos forman la sinonimia
taxa plural de taxón
taxon (taxón) una determinada unidad de clasificación, p.e.: género, especie, subespecie
var. variedad

* Expresamos nuestro agradecimiento al Dr. Aaron Davis, Real Jardín Botánico de Kew, por la presentación de esta guía.

11. Bibliografía

Principales fuentes de referencia utilizadas para la recopilación de las listas:

Ball, J.S. (1978). *Southern African Epiphytic Orchids*. Johannesburg: Conservation Press. pp248.

Bechtel, H., Cribb, P.J. & Launert, G.O.E (1992). *The Manual of Cultivated Orchid Species*. 3rd ed. London: Blandford. pp 585.

Cadet, J. (1989). *Joyaux de nos Forets: Les Orchidees de la Reunion*. Saint Denis: Ile de la Reunion: Nouvelle Impr. Dionysienne

Carnevali, G. & Ramírez, I.M. (1999). *Brassavola* In Carnevali, G., Ramírez, I.M., Romero, G. & Vargas, C. (1999). *Orchidaceae* in Steyermark, J.A., Berry, P.E. & Holst, B.K. (eds.) *Flora of the Venezuelan Guyana*. 5. St Louis: Missouri Botanical Garden. pp833.

Lang Kaiyong, Chen Singchi, Luo Yibo & Zhu Guanghua (1999). *Flora Reipublicae Popularis Sinicae. Tomus 17: Angiospermae: Monocotyledoneae: Orchidaceae (1)*. Beijing: Science Press. pp463.

Chen Singchi, Tsi Zhanhuo, Lang Kaiyong & Zhu Guanghua (1999). *Flora Reipublicae Popularis Sinicae. Tomus 18: Angiospermae: Monocotyledoneae: Orchidaceae (2)*. Beijing: Science Press. pp551.

Chen Singchi, Tsi Zhanhuo, & Zhu Guanghua (1999). *Flora Reipublicae Popularis Sinicae. Tomus 19: Angiospermae: Monocotyledoneae: Orchidaceae (3)*. Beijing: Science Press. pp485.

Clements, M.A. (1982). *Preliminary Checklist of Australian Orchidaceae*. Canberra, Australia: National Botanic Gardens. pp216.

Colombian Orchid Society (1990). *Native Colombian Orchids: Volume 1 Acacallis - Dryadella*. Medellin, Colombia: Compania Litografica National S.A.

Colombian Orchid Society (1991). Native Colombian Orchids: Volume 3 Maxillaria - Ponthieva. Medellin, Colombia: Compania Litografica National S.A.

Comber, J.B. (1990). *Orchids of Java*. Bentham-Moxon Trust, Royal Botanic Gardens, Kew, UK. pp407.

Cribb, P.J. & Robbins, S. (1991). The Genus Bletilla in Cultivation. *Orchid Review 99*, 406-409.

Cribb, P.J. (1984). *Flora of Tropical East Africa: Orchidaceae 3*. Balkema, Rotterdam.

Dockrill, A.W. (1992). *Australian Indigenous Orchids Volume 1: The Epiphytes, the tropical terrestrial species*. Chipping Norton, NSW: Surrey Beatty and Sons. pp524.

Dressler, R.L. (1980). *Orchids of Panama*. Missouri Botanical Garden, USA.

Dunsterville, G.C.K. & Garay, L.A. (1976). *Venezuelan Orchids Illustrated*. 6. A.Deutsch, London.

Dunsterville, G.C.K. (1987). *Venezuelan Orchids*. Armitano, Venezuela.

DuPuy, D. et al. (1999). *The Orchids of Madagascar: annotated checklist*. Royal Botanic Gardens, Kew, UK. pp376.

Grove, D.L. (1995). *Vandas & Ascocendas & their combinations with other genera*. Portland, Oregon: Timber Press. pp241.

Karasawa, K. (1998). *Orchid Atlas Volume 5: Calanthe - Coelogyne*. Orchid Atlas Publishing Society, Tokyo.

La Croix, I. & E. (1997). *African Orchids in the Wild and Cultivation*. Portland, Oregon: Timber Press. pp379.

La Croix, I. et al. (1991). *Orchids of Malawi: the epiphytic and terrestrial orchids from South and East Central Africa*. Rotterdam: A.A. Balkema. pp358.

Pabst, G.F.J. & Dungs, F. (1975). *Orchidaceae Brasilienses*. Ed. 1. Hildesheim: Brucke-Verlag Schmersow.

Perrier de la Bathie, H. (1981). *Flora of Madagascar (Vascular Plants): 49th family Orchids*. Lodi, California: S.D. Beckman. pp542.

Richard, A. H. (1974). *Flora of the Lesser Antilles: Leeward and Windward Islands*. Jamaica Plain, Mass.: Arnold Arboretum, Harvard University.

Seidenfaden, G. & Wood, J.J. (1992). *The Orchids of Peninsular Malaysia and Singapore - A Revision of R.E. Holttum: Orchds of Malaya*. Fredensborg, Denmark: Olsen & Olsen.

Seidenfaden, G. (1975). Orchid Genera in Thailand I - III. *Dansk Botanisk Arkiv* 33(3). 1-228.

Seidenfaden, G. (1992). The Orchids of Indochina. *Opera Botanica* 114, Copenhagen.

Seidenfaden, G. (1988). Orchid Genera in Thailand XIV. Fifty-nine vandoid Genera. *Opera Botanica* 95, Copenhagen.

Senghas, K. (1997). *Miltonia und verwandte Gattungen*. Schweizerische Orchideen Gesellschaft, Zurich. pp119.

Stewart, J. & Campbell, B. (1996). *Orchids of Kenya.* Winchester: St Paul's Bibliographies,. pp176.

Stewart, J. (1979). Revision of the African species of Aerangis (Orchidacea). *Kew Bulletin* 34: 2.

Wiard, L.A. (1987). *An Introduction to the Orchids of Mexico.* Ithaca, N.Y: Comstock Pub. Assocs,. pp239.

Williams, L.O. (1951). The Orchidaceae of Mexico. *CEIBA*, 2.

Williams, L.O. (1956). An enumeration of the Orchidaceae of Central America, British Honduras & Panama. *CEIBA*, 5.

Wood, J.J. & Cribb, P.J. (1994). *A Checklist of the Orchids of Borneo.* Royal Botanic Gardens, Kew, UK. pp409.

PART I: ORCHIDACEAE BINOMIALS IN CURRENT USE
Ordered alphabetically on All Names for the genera:

Aerangis, *Angraecum*, *Ascocentrum*, *Bletilla*, *Brassavola*, *Calanthe*, *Catasetum*, *Miltonia*, *Miltonioides*, *Miltoniopsis*, *Renanthera*, *Renantherella*, *Rhynchostylis*, *Rossioglossum*, *Vanda* and *Vandopsis*

PREMIERE PARTIE: BINOMES D'ORCHIDACEAE ACTUELLEMENT EN USAGE
Par ordre alphabétique de tous le noms pour les genre:

Aerangis, *Angraecum*, *Ascocentrum*, *Bletilla*, *Brassavola*, *Calanthe*, *Catasetum*, *Miltonia*, *Miltonioides*, *Miltoniopsis*, *Renanthera*, *Renantherella*, *Rhynchostylis*, *Rossioglossum*, *Vanda* et *Vandopsis*

PARTE I: ORCHIDACEAE BINOMIALES UTILIZADOS NORMALMENTE
Presentados por orden alfabético: todos los nombres para el genero:

Aerangis, *Angraecum*, *Ascocentrum*, *Bletilla*, *Brassavola*, *Calanthe*, *Catasetum*, *Miltonia*, *Miltonioides*, *Miltoniopsis*, *Renanthera*, *Renantherella*, *Rhynchostylis*, *Rossioglossum*, *Vanda* y *Vandopsis*

Part I: All Names / Tous les Noms / Todos los Nombres

ALPHABETICAL LISTING OF ALL NAMES FOR THE GENERA:
Aerangis, Angraecum, Ascocentrum, Bletilla, Brassavola, Calanthe, Catasetum, Miltonia, Miltonioides, Miltoniopsis, Renanthera, Renantherella, Rhynchostylis, Rossioglossum, Vanda and *Vandopsis*

LISTES ALPHABETIQUES DE TOUS LES NOMS POUR LES GENRE:
Aerangis, Angraecum, Ascocentrum, Bletilla, Brassavola, Calanthe, Catasetum, Miltonia, Miltonioides, Miltoniopsis, Renanthera, Renantherella, Rhynchostylis, Rossioglossum, Vanda et *Vandopsis*

PRESENTACION POR ORDEN ALFABETICO DE TODOS LOS NOMBRES PARA EL GENERO:
Aerangis, Angraecum, Ascocentrum, Bletilla, Brassavola, Calanthe, Catasetum, Miltonia, Miltonioides, Miltoniopsis, Renanthera, Renantherella, Rhynchostylis, Rossioglossum, Vanda y *Vandopsis*

ALL NAMES TOUS LES NOMS TODOS LOS NOMBRES	ACCEPTED NAME NOM ACCEPTÉ NOMBRES ACEPTADOS
Aerangis alata	**Aerangis ellisii**
Aerangis albido-rubra	**Aerangis luteo-alba** var. **rhodosticta**
Aerangis alcicornis	
Aerangis anjoanensis	**Aerangis mooreana**
Aerangis appendiculata	
Aerangis arachnopus	
Aerangis articulata	
Aerangis avicularia	**Aerangis rostellaris**
Aerangis batesii	**Aerangis arachnopus**
Aerangis biloba	
Aerangis biloba ssp. *kirkii*	**Aerangis kirkii**
Aerangis biloboides	**Aerangis arachnopus**
Aerangis bouarensis	
Aerangis brachycarpa	
Aerangis brachyceras	**Cribbia brachyceras**
Aerangis buchlohii	**Aerangis rostellaris**
Aerangis buyssonii	**Aerangis ellisii**
Aerangis calantha	
Aerangis calligera	**Aerangis articulata**
Aerangis calodictyon	**Aerangis alcicornis**
Aerangis campyloplectron	**Aerangis biloba**
Aerangis carnea	
Aerangis carusiana	**Aerangis brachycarpa**
Aerangis caulescens	**Aerangis ellisii**
Aerangis citrata	
Aerangis clavigera	**Aerangis macrocentra**
Aerangis collum-cygni	
Aerangis compta	**Aerangis collum-cygni**
Aerangis concavipetala	
Aerangis confusa	
Aerangis cordatighandula	**Rangaeris rhipsalisocia**
Aerangis coriacea	
Aerangis crassipes	**Aerangis modesta**
Aerangis cryptodon* (Rchb.f.) Schltr.	
Aerangis cryptodon sensu Perrier	**Aerangis ellisii**
Aerangis curnowiana	

***For explanation see page 3, point 6**
***Voir les explications page 12, point 6**
***Para mayor explicación, véase la página 22, point 6**

ALL NAMES	ACCEPTED NAME
Aerangis decaryana	
Aerangis distincta	
Aerangis elegans	**Aerangis flexuosa**
Aerangis ellisii	
Aerangis ellisii var. **grandiflora**	
Aerangis englerianum	**Rangaeris muscicola**
Aerangis erythrurum	**Aerangis verdickii**
Aerangis falcifolia	**Rangaeris muscicola**
Aerangis fastuosa	
Aerangis fastuosa ssp. *angustifolia*	**Aerangis modesta**
Aerangis fastuosa ssp. *francoisii*	**Aerangis fastuosa**
Aerangis fastuosa ssp. *grandidieri*	**Aerangis fastuosa**
Aerangis fastuosa ssp. *maculata*	**Aerangis fastuosa**
Aerangis fastuosa ssp. *rotundifolia*	**Aerangis fastuosa**
Aerangis fastuosa ssp. *vondrozensis*	**Aerangis fastuosa**
Aerangis filipes	**Rhipidoglossum curvatum**
Aerangis flabellifolia	**Aerangis brachycarpa**
Aerangis flexuosa	
Aerangis floribunda	**Rangaeris muscicola**
Aerangis friesiorum Schltr.	**Aerangis thomsonii***
Aerangis friesiorum sensu Tweedie non Schltr.	**Aerangis brachycarpa**
Aerangis fuscata* (Rchb.f.) Schltr.	
Aerangis fuscata sensu Perrier	**Aerangis stylosa**
Aerangis gracillima	
Aerangis graminifolia	**Ypsilopus longifolius**
Aerangis grantii	**Aerangis kotschyana**
Aerangis gravenreuthii	
Aerangis henriquesiana	**Aerangis flexuosa**
Aerangis hologlottis	
Aerangis hyaloides	
Aerangis ikopana	**Aerangis mooreana**
Aerangis jacksonii	
Aerangis kirkii	
Aerangis kotschyana (Rchb.f.) Schltr.	
Aerangis kotschyana sensu Morris non Schltr.	**Aerangis splendida**
Aerangis kotschyana sensu Schelpe non Schltr.	**Aerangis verdickii***
Aerangis kotschyi	**Aerangis kotschyana**
Aerangis laurentii	**Summerhayesia laurentii**
Aerangis lutambae	**Aerangis alcicornis**
Aerangis luteo-alba	
Aerangis luteo-alba var. **luteo-alba**	
Aerangis luteo-alba var. **rhodosticta**	
Aerangis macrocentra	
Aerangis maireae	
Aerangis malmquistiana	**Aerangis cryptodon**
Aerangis megaphylla	
Aerangis mixta	**Rangaeris muscicola**
Aerangis moandensis	**Angraecum moandensis**
Aerangis modesta	
Aerangis monantha	**Aerangis fuscata**
Aerangis montana	
Aerangis mooreana	
Aerangis muscicola	**Rangaeris muscicola**
Aerangis mystacidii	
Aerangis mystacidioides	**Aerangis mystacidii**
Aerangis oligantha	
Aerangis pachyura sensu Morris non Schltr.	**Aerangis appendiculata**

***For explanation see page 3, point 6**
***Voir les explications page 12, point 6**
***Para mayor explicación, véase la página 22, point 6**

Part I: All Names / Tous les Noms / Todos los Nombres

ALL NAMES	ACCEPTED NAME
Aerangis pachyura (Rolfe) Schltr.	**Aerangis mystacidii***
Aerangis pallida	
Aerangis pallidiflora	
Aerangis parvula	**Aerangis calantha**
Aerangis phalaenopsis	**Aerangis megaphylla**
Aerangis platyphylla	**Aerangis ellisii**
Aerangis potamophila	**Angraecum potamophilum**
Aerangis primulina	**Aerangis × primulina**
Aerangis × primulina	
Aerangis pulchella	
Aerangis pumilio	**Aerangis hyaloides**
Aerangis punctata	
Aerangis rhipsalisocia	**Rangaeris rhipsalisocia**
Aerangis rhodosticta	**Aerangis luteo-alba** var. **rhodosticta**
Aerangis rohlfsiana	**Aerangis brachycarpa**
Aerangis roseocalcarata	**Aerangis calantha**
Aerangis rostellaris	
Aerangis rusituensis	**Aerangis verdickii** ssp. **rusituensis**
Aerangis sankuruensis	**Aerangis calantha**
Aerangis schliebenii	**Aerangis verdickii**
Aerangis seegeri	**Aerangis pallidiflora**
Aerangis solheidii	**Rangaeris muscicola**
Aerangis somalensis	
Aerangis spiculata	
Aerangis splendida	
Aerangis stelligera	
Aerangis stylosa (Rolfe) Schltr.	
Aerangis stylosa sensu Perrier	**Aerangis articulata***
Aerangis thomsonii	
Aerangis ugandensis	
Aerangis umbonata	**Aerangis fuscata**
Aerangis venusta	**Aerangis articulata**
Aerangis verdickii	
Aerangis verdickii ssp. **rusituensis**	
Aeranthes englerianus	**Angraecum kraenzlinianum**
Aeranthes gladiifolius	**Angraecum mauritianum**
Aeranthes leonis	**Angraecum leonis**
Aeranthes pectinatus	**Angraecum pectinatum**
Aeranthes sesquipedalis	**Angraecum sesquipedale**
Aeranthes thouarsii	**Angraecum filicornu**
Aeranthes trichoplectron	**Angraecum trichoplectron**
Aeranthus calceolus	**Angraecum calceolus**
Aeranthus curnowianus	**Angraecum curnowianum**
Aeranthus gladiifolius	**Angraecum mauritianum**
Aeranthus gravenreuthii	**Aerangis gravenreuthii**
Aeranthus leonis	**Angraecum leonis**
Aeranthus meirax	**Angraecum meirax**
Aerides ampullacea	**Ascocentrum ampullaceum**
Aerides coriaceum	**Angraecum coriaceum**
Aerides elongata	**Renanthera elongata**
Aerides guttata	**Rhynchostylis retusa**
Aerides maculatum	**Vanda spathulata**
Aerides matutina	**Renanthera matutina**
Aerides orthocentra	**Vanda testacea**
Aerides praemorsa	**Rhynchostylis retusa**
Aerides retusum	**Rhynchostylis retusa**

***For explanation see page 3, point 6**
***Voir les explications page 12, point 6**
***Para mayor explicación, véase la página 22, point 6**

ALL NAMES	ACCEPTED NAME
Aerides spicata	**Rhynchostylis retusa**
Aerides sulingi	**Renanthera sulingi**
Aerides tesselatum Wight in Wall.	**Vanda spathulata***
Aerides tesselatum Thw. non Wight	**Vanda thwaitesii**
Aerides testacea	**Vanda testacea**
Aerides undulatum	**Rhynchostylis retusa**
Aerides wightiana auct. non Lindl.	**Vanda lilacina***
Aerides wightiana Lindl.	**Vanda testacea**
Aerobion calceolus	**Angraecum calceolus**
Aerobion caulescens	**Angraecum caulescens**
Aerobion citratum	**Aerangis citrata**
Aerobion crassum	**Angraecum crassum**
Aerobion cucullatum	**Angraecum cucullatum**
Aerobion filicornu	**Angraecum filicornu**
Aerobion gladiifolium	**Angraecum mauritianum**
Aerobion implicatum	**Angraecum implicatum**
Aerobion inapertum	**Angraecum inapertum**
Aerobion multiflorum	**Angraecum multiflorum**
Aerobion palmiforme	**Angraecum palmiforme**
Aerobion pectinatum	**Angraecum pectinatum**
Aerobion striatum	**Angraecum striatum**
Aerobion superbum	**Angraecum eburneum** ssp. **superbum**
Aerobion triquetrum	**Angraecum triquetrum**
Alismorchis abbreviata	**Calanthe abbreviata**
Alismorchis angusta	**Calanthe angusta**
Alimorchis masuca	**Calanthe sylvatica**
Alismorchis parviflora	**Calanthe flava**
Alismorchis zollingeri	**Calanthe ceciliae**
Alismographis lyroglossa	**Calanthe discolor** ssp. **discolor**
Alismorkis angusta	**Calanthe odora**
Alismorkis angustifolia	**Calanthe angustifolia**
Alismorkis centrosis	**Calanthe sylvatica**
Alismorkis diploxiphion	**Calanthe triplicata**
Alismorkis discolor	**Calanthe discolor** ssp. **discolor**
Alismorkis foerstermannii	**Calanthe lyroglossa**
Alismorkis furcata	**Calanthe triplicata**
Alismorkis gracillima	**Calanthe triplicata**
Alismorkis japonica	**Calanthe alismaefolia**
Alismorkis lyroglossa	**Calanthe lyroglossa**
Alismorkis phajoides	**Calanthe angustifolia**
Alismorkis pleiochroma	**Calanthe sylvatica**
Alismorkis pulchra	**Calanthe pulchra**
Alismorkis reflexa	**Calanthe reflexa**
Alismorkis rosea	**Calanthe rosea**
Alismorkis textori	**Calanthe sylvatica**
Alismorkis veratrifolia	**Calanthe triplicata**
Amblyglottis abbreviata	**Calanthe abbreviata**
Amblyglottis angustifolia	**Calanthe angustifolia**
Amblyglottis emarginata	**Calanthe sylvatica**
Amblyglottis flava	**Calanthe flava**
Amblyglottis pilosa	**Calanthe vestita**
Amblyglottis pulchra	**Calanthe pulchra**
Amblyglottis speciosa	**Calanthe speciosa**
Amblyglottis veratrifolia Blume	**Calanthe ceciliae**
Amblyglottis veratrifolia (Willd.) Blume	**Calanthe triplicata***
Angorchis alcicornis	**Aerangis alcicornis**
Angorchis arachnopus	**Aerangis arachnopus**

***For explanation see page 3, point 6**
***Voir les explications page 12, point 6**
***Para mayor explicación, véase la página 22, point 6**

Part I: All Names / Tous les Noms / Todos los Nombres

ALL NAMES	ACCEPTED NAME
Angorchis articulata	**Aerangis articulata**
Angorchis biloba	**Aerangis biloba**
Angorchis brongniartiana	**Angraecum eburneum** ssp. **superbum**
Angorchis campyloplectron	**Aerangis biloba**
Angorchis citrata	**Aerangis citrata**
Angorchis clavigera	**Angraecum clavigerum**
Angorchis conchifera	**Angraecum conchiferum**
Angorchis crassa	**Angraecum crassum**
Angorchis cryptodon	**Aerangis cryptodon**
Angorchis cucullata	**Angraecum cucullatum**
Angorchis curnowiana	**Angraecum curnowianum**
Angorchis eburnea	**Angraecum eburneum**
Angorchis ellisii	**Aerangis ellisii**
Angorchis fastuosa	**Aerangis fastuosa**
Angorchis flabellifolia	**Aerangis brachycarpa**
Angorchis fragrans	**Angraecum cucullatum**
Angorchis gladiifolia	**Angraecum mauritianum**
Angorchis hyaloides	**Aerangis hyaloides**
Angorchis implicata	**Angraecum implicatum**
Angorchis infundibularis	**Angraecum infundibulare**
Angorchis modesta	**Aerangis modesta**
Angorchis palmiformis	**Angraecum palmiforme**
Angorchis parvula	**Angraecum parvulum**
Angorchis pectangis	**Angraecum pectinatum**
Angorchis pectinata	**Angraecum pectinatum**
Angorchis pusilla	**Angraecum pusillum**
Angorchis rhodosticta	**Aerangis luteo-alba** var. **rhodosticta**
Angorchis rostrata	**Angraecum rostratum**
Angorchis scottiana	**Angraecum scottianum**
Angorchis sesquipedalis	**Angraecum sesquipedale**
Angorchis striata	**Angraecum striatum**
Angorchis superba	**Angraecum eburneum** ssp. **superbum**
Angorchis teretifolia	**Angraecum teretifolium**
Angraecum abietinum	**Angraecum humblotianum**
Angraecum acutimarginatum	**Tridactyle inaequilonga**
Angraecum acuto-emarginatum	**Tridactyle inaequilonga**
Angraecum acutipetalum	
Angraecum acutipetalum ssp. **analabeensis**	
Angraecum acutipetalum ssp. **ankeranae**	
Angraecum acutum	**Diaphananthe acuta**
Angraecum affine	
Angraecum albido-rubrum	**Aerangis luteo-alba** var. **rhodosticta**
Angraecum alcicorne	**Aerangis alcicornis**
Angraecum alleizettei	
Angraecum aloifolium	
Angraecum althoffii	**Diaphananthe pellucida**
Angraecum amaniense	**Angraecopsis tenerrima**
Angraecum ambongoense	**Lemurella ambongensis**
Angraecum ambrense	
Angraecum amplexicaule	
Angraecum ampullaceum	
Angraecum andasibeense	
Angraecum andersonii	**Microcoelia caespitosa**
Angraecum andringitranum	
Angraecum angustifolium	**Cyrtorchis aschersonii**
Angraecum angustipetalum	
Angraecum angustum (Rolfe) Summerh.	

***For explanation see page 3, point 6**
***Voir les explications page 12, point 6**
***Para mayor explicación, véase la página 22, point 6**

ALL NAMES	ACCEPTED NAME
Angraecum angustum Rolfe	**Aerangis verdickii***
Angraecum anjouanense	**Jumellea anjouanensis**
Angraecum ankeranense	
Angraecum anocentrum	**Angraecum calceolus**
Angraecum antennatum	**Cyrtorchis monteiroae**
Angraecum aphyllum	**Solenangis aphylla**
Angraecum apiculatum	**Aerangis biloba**
Angraecum apiculatum ssp. *kirkii*	**Aerangis kirkii**
Angraecum aporoides	
Angraecum appendiculatum	**Bonniera appendiculata**
Angraecum appendiculoides	
Angraecum arachnites	**Angraecum germinyanum**
Angraecum arachnopus	**Aerangis arachnopus**
Angraecum armeniacum	**Tridactyle armeniaca**
Angraecum arnoldianum	**Angraecum eichlerianum**
Angraecum arthrophyllum	**Angraecum pungens**
Angraecum articulatum	**Aerangis articulata**
Angraecum aschersonii	**Cyrtorchis monteiroae**
Angraecum ashantense	**Listrostachys ashautensis**
Angraecum astroarche	
Angraecum augustum	**Aerangis verdickii**
Angraecum aviceps	
Angraecum avicularium	**Aerangis rostellaris**
Angraecum bakeri	**Diaphananthe bidens**
Angraecum bancoense	
Angraecum baronii	
Angraecum baronii	**Angraecum costatum**
Angraecum batesii	**Aerangis arachnopus**
Angraecum bathiei	**Angraecum germinyanum**
Angraecum bathiei ssp. *peracuminatum*	**Angraecum germinyanum**
Angraecum bemarivoense	
Angraecum bicallosum	
Angraecum bicaudatum	**Tridactyle bicaudata**
Angraecum bidens	**Diaphananthe bidens**
Angraecum bieleri	**Microcoelia bieleri**
Angraecum biloboides	**Aerangis arachnopus**
Angraecum bilobum Lindl.	**Aerangis biloba**
Angraecum bilobum sensu Engl. non Lindl	**Aerangis brachycarpa**
Angraecum bilobum Schweinf. non Lindl.	**Aerangis brachycarpa**
Angraecum bilobum ssp. *kirkii*	**Aerangis kirkii**
Angraecum birrimense	
Angraecum bistortum	**Cyrtorchis ringens**
Angraecum bokoyense	**Calyptrochilum christyanum**
Angraecum bolusii	**Tridactyle tridentata**
Angraecum boonei	**Angraecum angustipetalum**
Angraecum borbonicum	
Angraecum bosseri	**Angraecum sesquipedale** var. **angustifolium**
Angraecum boutoni	**Angraecopsis pobeguinii**
Angraecum brachycarpum	**Aerangis brachycarpa**
Angraecum brachyrhopalon	
Angraecum bracteosum	
Angraecum braunii	**Angraecum viride**
Angraecum breve	
Angraecum brevicornu	
Angraecum brevifolium	**Campylocentrum micranthum**
Angraecum brongniartianum	**Angraecum eburneum** ssp. **superbum**
Angraecum brunneo-maculatum	**Ancistrorhynchus clandestinus**

***For explanation see page 3, point 6**
***Voir les explications page 12, point 6**
***Para mayor explicación, véase la página 22, point 6**

Part I: All Names / Tous les Noms / Todos los Nombres

ALL NAMES	ACCEPTED NAME
Angraecum buchholzianum	**Cyrtorchis ringens**
Angraecum bueae	**Diaphananthe bueae**
Angraecum burchellii	**Angraecum pusillum**
Angraecum buyssonii	**Aerangis ellisii**
Angraecum cadetii	
Angraecum caespitosum	**Microcoelia caespitosa**
Angraecum caffrum	**Margelliantha caffra**
Angraecum calanthum	**Aerangis calantha**
Angraecum calceolus	
Angraecum calligerum	**Aerangis articulata**
Angraecum campyloplectron Rchb.f. sensu Stewart	**Aerangis biloba**
Angraecum campyloplectron Rchb.f. sensu Seidenf.	**Ascocentrum ampullaceum***
Angraecum canaliculatum	**Angraecum subulatum**
Angraecum capense	**Mystacidium capense**
Angraecum capitatum	**Ancistrorhynchus capitatus**
Angraecum caricifolium	
Angraecum carifolium	**Angraecum caricifolium**
Angraecum carinatum	**Eulophia virens**
Angraecum carpophorum	**Angraecum calceolus**
Angraecum carusianum	**Aerangis brachycarpa**
Angraecum catati	**Angraecum rutenbergianum**
Angraecum caudatum	**Plectrelminthus caudatus**
Angraecum caulescens	
Angraecum caulescens var. *multiflorum*	**Angraecum multiflorum**
Angraecum cephalotes	**Ancistrorhynchus mettemiae**
Angraecum chaetopodum	
Angraecum chailluanum	**Cyrtorchis chailluana**
Angraecum chamaeanthus	
Angraecum chermezoni	
Angraecum chevalieri	**Angraecum moandense**
Angraecum chiloschistae	**Microcoelia exilis**
Angraecum chimanimaniense	
Angraecum chloranthum	
Angraecum christyanum	**Calyptrochilum christyanum**
Angraecum cilaosianum	
Angraecum citratum	**Aerangis citrata**
Angraecum claessensii	
Angraecum clandestinum	**Ancistrorhynchus clandestinus**
Angraecum clandestinum ssp. *durandianum*	**Ancistrorhynchus clandestinus**
Angraecum clandestinum ssp. *stenophyllum*	**Ancistrorhynchus clandestinus**
Angraecum clavatum (Rendle) Schltr.	**Angraecum multinominatum**
Angraecum clavatum sensu Orstom	**Solenangis clavata**
Angraecum clavigerum	
Angraecum comorense Kraenzl. non (Rchb.f.) Finet	**Angraecum eburneum** ssp. **superbum***
Angraecum comorense fide Rice	**Jumellea comorensis**
Angraecum compactum	
Angraecum compressicaule	
Angraecum conchiferum	
Angraecum conchoglossum	**Angraecum germinyanum**
Angraecum confusa	**Jumellea confusa**
Angraecum conicum	**Solenangis conica**
Angraecum cordatiglandulum	**Rangaeris rhipsalisocia**
Angraecum cordemoyi	
Angraecum coriaceum	
Angraecum cornigerum	
Angraecum cornucopiae	
Angraecum cornutum	**Solenangis cornuta**

***For explanation see page 3, point 6**
***Voir les explications page 12, point 6**
***Para mayor explicación, véase la página 22, point 6**

ALL NAMES	ACCEPTED NAME
Angraecum corynoceras	
Angraecum costatum	
Angraecum coutrixii	
Angraecum cowanii	**Jumellea cowanii**
Angraecum crassiflorum	**Angraecum crassum**
Angraecum crassifolia	**Angraecum sacciferum**
Angraecum crassifolium	
Angraecum crassum	
Angraecum crenatum	**Cyrtorchis crenata**
Angraecum cribbianum	
Angraecum crinale	**Microcoelia caespitosa**
Angraecum cryptodon	**Aerangis cryptodon**
Angraecum crystallinum	**Diaphananthe bidens**
Angraecum cucullatum	
Angraecum cufodontii	**Rangaeris amaniensis**
Angraecum culiciferum	**Lemurella culicifera**
Angraecum cultriforme	
Angraecum curnowianum	
Angraecum curvatum	**Diaphananthe curvata**
Angraecum curvicalcar	
Angraecum curvicaule	
Angraecum curvipes	
Angraecum cyclochilum	**Solenangis cornuta**
Angraecum dactyloceras	**Podangis dactyloceras**
Angraecum danguyanum	
Angraecum dasycarpum	
Angraecum dauphinense	
Angraecum decaryanum	
Angraecum decipiens	
Angraecum deflexicalcaratum	**Chauliodon deflexicalcarata**
Angraecum defoliatum	**Solenangis aphylla**
Angraecum dendrobiopsis	
Angraecum descendens	**Aerangis articulata**
Angraecum dichaeoides	**Angraecum baronii**
Angraecum didieri	
Angraecum distichophyllum	**Angraecum striatum**
Angraecum distichum	
Angraecum distichum ssp. *grandifolium*	**Angraecum aporoides**
Angraecum divaricatum	
Angraecum dives	
Angraecum divitiflorum	**Microterangis boutonii**
Angraecum dolabriforme	**Angraecopsis dolabriformis**
Angraecum dolichorrhizum	**Microcoelia dolichorrhiza**
Angraecum dollii	
Angraecum doratophyllum	
Angraecum dorotheae	**Diaphananthe dorotheae**
Angraecum drouhardii	
Angraecum dryadum	
Angraecum dubuyssonii	**Aerangis ellisii**
Angraecum durandinum	**Ancistrorhynchus clandestinus**
Angraecum ealaense	**Cyrtorchis injoloensis**
Angraecum eburneum	
Angraecum eburneum ssp. *brongniartianum*	**Angraecum eburneum** ssp. **superbum**
Angraecum eburneum ssp. **giryamae**	
Angraecum eburneum ssp. *longicalcar*	**Angraecum eburneum** var. **longicalcar**
Angraecum eburneum ssp. **superbum**	
Angraecum eburneum ssp. *typicum*	**Angraecum eburneum**

***For explanation see page 3, point 6**
***Voir les explications page 12, point 6**
***Para mayor explicación, véase la página 22, point 6**

Part I: All Names / Tous les Noms / Todos los Nombres

ALL NAMES	ACCEPTED NAME
Angraecum eburneum ssp. *virens*	**Angraecum eburneum**
Angraecum eburneum ssp. **xerophilum**	
Angraecum eburneum var. **longicalcar**	
Angraecum egertonii	
Angraecum eichlerianum	
Angraecum eichlerianum var. **curvicalcaratum**	
Angraecum eigeltii	**Campylocentrum fasciola**
Angraecum elatum	**Cryptopus elatus**
Angraecum elegans	**Aerangis flexuosa**
Angraecum elephantinum	
Angraecum elliotii	
Angraecum ellisii	**Aerangis ellisii**
Angraecum ellisii var. *occidentale*	**Aerangis gravenreuthii**
Angraecum emarginatum	**Calyptrochilum emarginatum**
Angraecum englerianum (Kraenzl.) Schltr. non Kraenzl.	**Angraecum kraenzlinianum**
Angraecum englerianum sensu Summerh.	**Rangaeris muscicola***
Angraecum equitans	
Angraecum erecto-calcaratum	**Diaphananthe rutila**
Angraecum erectum	
Angraecum erythrurum	**Aerangis verdickii**
Angraecum evrardianum	
Angraecum exile	**Jumellea exilis**
Angraecum expansum	
Angraecum expansum ssp. **inflatum**	
Angraecum falcatum	**Neofinetia falcatum**
Angraecum falcifolium	
Angraecum fasciola	**Campylocentrum fasciola**
Angraecum fastuosum	**Aerangis fastuosa**
Angraecum ferkoanum	
Angraecum filicornoides	**Jumellea filicornoides**
Angraecum filicornu* Thouars	
Angraecum filicornu sensu Kraenzl. non Thouars	**Angraecum sterrophyllum**
Angraecum filifolium	**Tridactyle tridentata**
Angraecum filiforme Lindl.	**Campylocentrum filiforme**
Angraecum filiforme Schltr.	**Nephrangis filiformis**
Angraecum filipes	**Rhipidoglossum curvatum**
Angraecum fimbriatum	**Tridactyle bicaudata**
Angraecum fimbritipetalum	**Tridactyle fimbritipetalum**
Angraecum finetianum	**Angraecum humblotianum**
Angraecum firthii	
Angraecum flabellifolium	**Aerangis brachycarpa**
Angraecum flanaganii	**Mystacidium flanaganii**
Angraecum flavidum	
Angraecum flexuosum	**Aerangis flexuosa**
Angraecum floribundum	
Angraecum florulentum	
Angraecum forcipatum	**Podangis dactyloceras**
Angraecum fournerianum	**Sobennikoffia fourneriana**
Angraecum fournierae	**Aerangis stylosa**
Angraecum foxii	**Angraecum rhynchoglossum**
Angraecum fragrans	**Jumellea fragrans**
Angraecum frommianum	**Tridactyle tricuspis**
Angraecum funale	**Dendrophylax funalis**
Angraecum furvum	**Vanda concolor**
Angraecum fuscatum Rchb.f.	**Aerangis fuscata***
Angraecum fuscatum sensu Carriere	**Aerangis spiculata**
Angraecum gabonense	

***For explanation see page 3, point 6**
***Voir les explications page 12, point 6**
***Para mayor explicación, véase la página 22, point 6**

ALL NAMES	ACCEPTED NAME
Angraecum galeandrae	**Eurychone galeandrae**
Angraecum geniculatum	
Angraecum gentilii	**Tridactyle gentilii**
Angraecum germinyanum	
Angraecum gerrardii	**Diaphananthe xanthopollinia**
Angraecum gilpinae	**Microcoelia gilpinae**
Angraecum giryamae	**Angraecum eburneum** ssp. **giryamae**
Angraecum gladiifolium	**Angraecum mauritianum**
Angraecum globuloso-calcaratum	**Diaphananthe globulosocalcara**
Angraecum globulosum	**Microcoelia globulosa**
Angraecum glomeratum	**Ancistrorhynchus cephalotes**
Angraecum goetzeanum	**Tridactyle tridentata**
Angraecum gracilipes	**Jumellea sagittata**
Angraecum gracilis	**Chamaeangis gracilis**
Angraecum gracillimum	**Aerangis gracillima**
Angraecum gradidieranum	**Neobathiea gradidierana**
Angraecum graminifolium (Ridl.) Schltr.	**Angraecum pauciramosum***
Angraecum graminifolium (Kraenzl.) Engl.	**Ypsilopus longifolius**
Angraecum grandiflorum	**Aeranthus grandiflorus**
Angraecum grantii	**Aerangis kotschyana**
Angraecum gravenreutii	**Aerangis gravenreuthii**
Angraecum guillauminii	
Angraecum guyonianum	**Microcoelia globulosa**
Angraecum gyriamae	**Angraecum eburneum** ssp. **giryamae**
Angraecum henriquesianum Rolfe	**Aerangis flexuosa***
Angraecum henriquesianum Ridl.	**Cyrtorchis henriquesiana**
Angraecum hermannii	
Angraecum hildebrandtii	**Microterangis hildebrandtii**
Angraecum hislopii	**Tridactyle tridentata**
Angraecum hologlottis	**Aerangis hologlottis**
Angraecum humbertii	
Angraecum humblotianum	
Angraecum humblotii (Finet) Summerh.	**Angraecum humblotianum**
Angraecum humblotii Rchb.f. ex Rolfe	**Angraecum leonis***
Angraecum humile	
Angraecum huntleyoides	
Angraecum hyaloides	**Aerangis hyaloides**
Angraecum ichneumoneum	**Chamaeangis ichneumonea**
Angraecum imbricatum (Sw.) Schltr.	**Angraecum distichum**
Angraecum imbricatum Schltr.	**Epidendrum imbricatum**
Angraecum imbricatum Lindl.	**Calyptrochilum emarginatum***
Angraecum imerinense	
Angraecum implicatum	
Angraecum inaequilongum	**Tridactyle inaequilonga**
Angraecum inapertum	
Angraecum infundibulare	
Angraecum injoloense	**Cyrtorchis injoloensis**
Angraecum ischnopus Schltr. 1916	**Angraecopsis ischnopus**
Angraecum ischnopus Schltr. 1905	**Angraecum tenuipes**
Angraecum ivorense	**Calyptrochilum christyanum**
Angraecum jamaicense	**Campylocentrum micranthum**
Angraecum jumelleanum	**Jumellea jumelleana**
Angraecum kamerunense	**Diaphananthe kamerunensis**
Angraecum keniae	
Angraecum kimballianum	**Oeoniella polystachys**
Angraecum kindtianum	**Tridactyle tridactylites**
Angraecum kirkii	**Aerangis kirkii**

***For explanation see page 3, point 6**
***Voir les explications page 12, point 6**
***Para mayor explicación, véase la página 22, point 6**

Part I: All Names / Tous les Noms / Todos los Nombres

ALL NAMES	ACCEPTED NAME
Angraecum koehleri	**Microcoelia koehleri**
Angraecum konduense	**Microcoelia konduensis**
Angraecum kotschyanum	**Aerangis kotschyana**
Angraecum kotschyi	**Aerangis kotschyana**
Angraecum kraenzlinianum	
Angraecum laciniatum	**Tridactyle bicaudata**
Angraecum laggiarae	
Angraecum lagosense	**Tridactyle lagosensis**
Angraecum lansbergii	**Campylocentrum micranthum**
Angraecum latibracteatum	**Cyrtorchis brownii**
Angraecum laurentii	**Summerhayesia laurentii**
Angraecum lecomtei	
Angraecum ledermannianum	**Diaphananthe vandaeformis**
Angraecum leonii	**Angraecum leonis**
Angraecum leonis	
Angraecum lepidotum	**Tridactyle anthomaniaca**
Angraecum letouzeyi	
Angraecum lignosum	**Jumellea lignosa**
Angraecum ligulatum	**Angraecum affine**
Angraecum liliodorum	**Jumellea liliodora**
Angraecum lindenii	**Polyradicion lindenii**
Angraecum linearifolium Garay	
Angraecum linearifolium Cribb	**Angraecum umbrosum**
Angraecum lisowskianum	
Angraecum litorale	
Angraecum longicalcar	**Angraecum eburneum** var. **longicalcar**
Angraecum longicaule	
Angraecum longinode	
Angraecum lujaei	**Eurychone galeandrae**
Angraecum luridum	**Graphorkis lurida**
Angraecum luteo-album	**Aerangis luteo-alba**
Angraecum macilentum	
Angraecum macrocentrum	**Aerangis macrocentra**
Angraecum macrorrhynchium	**Microcoelia macrorrhynchia**
Angraecum maculatum	**Oeceoclades maculata**
Angraecum madagascariense	
Angraecum magdalenae	
Angraecum magdalenae var. **latilabellum**	
Angraecum mahavavense	
Angraecum maheense	**Angraecum zeylanicum**
Angraecum majale	**Jumellea majalis**
Angraecum malangeanum	**Calyptrochilum christyanum**
Angraecum marcrorhynchum	**Encheiridion marcrorhynchum**
Angraecum marii	
Angraecum marsupio-calcaratum	**Calyptrochilum christyanum**
Angraecum maudae	**Bolusiella maudiae**
Angraecum mauritianum	
Angraecum maxillarioides	**Jumellea maxillarioides**
Angraecum megalorrhizzum	**Microcoelia megalorrhiza**
Angraecum meirax	
Angraecum melanostictum	
Angraecum metallicum	
Angraecum micranthum	**Campylocentrum micranthum**
Angraecum microcharis	
Angraecum micropetalum	**Microcoelia caespitosa**
Angraecum microphyton	**Angraecum tenellum**
Angraecum minus	

ALL NAMES	ACCEPTED NAME
Angraecum minutissimum	
Angraecum minutum* Frapp. ex Cordem.	
Angraecum minutum A.Chev.	**Angraecum minutum**
Angraecum minutum A.Chev. ex Summerh.	**Diaphananthe curvata**
Angraecum mirabile Hort non Schltr.	**Aerangis luteo-alba** var. **rhodosticta**
Angraecum mirabile Schltr.	
Angraecum moandense	
Angraecum modestum	**Aerangis modesta**
Angraecum modicum	
Angraecum mofakoko	
Angraecum moloneyi	**Calyptrochilum christyanum**
Angraecum mombasense	**Calyptrochilum christyanum**
Angraecum monodon	**Diaphananthe bidens**
Angraecum monophyllum	**Oeceoclades maculata**
Angraecum montanum	**Diaphananthe montana**
Angraecum mooreanum	**Aerangis mooreana**
Angraecum moratii	
Angraecum muansae	**Diaphananthe fragrantissima**
Angraecum multiflorum	
Angraecum multinominatum	
Angraecum muriculatum	**Tridactyle muriculatum**
Angraecum muscicolum	
Angraecum musculiferum	
Angraecum myrianthum	
Angraecum mystacidii	**Aerangis mystacidii**
Angraecum nalaense	**Tridactyle nalaensis**
Angraecum nanum	
Angraecum nasutum	
Angraecum neglectum	**Angraecum triquetrum**
Angraecum neglectum ssp. *curtum*	**Angraecum triquetrum**
Angraecum neglectum ssp. *genuinum*	**Angraecum triquetrum**
Angraecum neglectum ssp. *longifolium*	**Angraecum triquetrum**
Angraecum nutans	**Jumellea nutans**
Angraecum nzoanum	
Angraecum obanense	**Diaphananthe obanensis**
Angraecum oberonia	
Angraecum obesum	
Angraecum oblongifolium	
Angraecum obversifolium	
Angraecum occidentale	**Angraecopsis tridens**
Angraecum ochraceum	
Angraecum odoratissimum	**Chamaeangis odoratissima**
Angraecum oeonioides	**Solenangis clavata**
Angraecum oliganthum	**Microterangis oligantha**
Angraecum onivense	
Angraecum ophioplectron	**Jumellea ophioplectron**
Angraecum organense	**Campylocentrum organense**
Angraecum ornithorrhynchium	**Campylocentrum ornithorrhynchium**
Angraecum ovalifolium	**Calyptrochilum christyanum**
Angraecum pachyurum Rolfe	**Aerangis mystacidii***
Angraecum pachyurum Kraenzl.	**Jumellea pachyra**
Angraecum pallidum	**Aerangis pallida**
Angraecum palmatum	**Angraecum palmiforme**
Angraecum palmicolum	
Angraecum palmiforme Thoars	
Angraecum palmiforme H.Perrier	**Angraecum linearifolium***
Angraecum panicifolium	

***For explanation see page 3, point 6**
***Voir les explications page 12, point 6**
***Para mayor explicación, véase la página 22, point 6**

Part I: All Names / Tous les Noms / Todos los Nombres

ALL NAMES	ACCEPTED NAME
Angraecum paniculatum	**Angraecum calceolus**
Angraecum parcum	**Angraecum sacciferum**
Angraecum parviflorum	**Angraecopsis parviflora**
Angraecum parvulum	
Angraecum patens	**Angraecum calceolus**
Angraecum pauciramosum	
Angraecum pectinatum	
Angraecum pellucidum	**Diaphananthe pellucida**
Angraecum penicillatum	**Jumellea penicillata**
Angraecum penzigianum	
Angraecum pergracile	
Angraecum perhumile	
Angraecum perparvulum	
Angraecum perrieri	**Microcoelia perrieri**
Angraecum pertusum	**Listrostachys pertusa**
Angraecum pescatorianum	**Listrostachys pertusa**
Angraecum petterssonianum	
Angraecum peyrotii	
Angraecum philippinense	**Amesiella philippinensis**
Angraecum physophorum	**Microcoelia physophora**
Angraecum pingue	
Angraecum pinifolium	
Angraecum plehnianum	**Diaphananthe plehniana**
Angraecum podochiloides	
Angraecum poeppigii	**Campylocentrum poeppigii**
Angraecum polystachyum Lindl.	**Campylocentrum polystachyum**
Angraecum polystachyum A.Rich.	**Oeoniella polystachys**
Angraecum poophyllum	**Angraecum pauciramosum**
Angraecum popowii	
Angraecum potamophilum	
Angraecum praestans	
Angraecum primulinum	**Aerangis × primulina**
Angraecum protensum	
Angraecum pseudodidieri	
Angraecum pseudofilicornu	
Angraecum pseudopetiolatum	
Angraecum pterophyllum	
Angraecum pugioniforme	**Cleisostoma pugioniforme**
Angraecum pulchellum	**Aerangis pulchella**
Angraecum pumilio	
Angraecum pungens	
Angraecum pusillum	
Angraecum pygmaeum	
Angraecum pynaertii	**Calyptrochilum christyanum**
Angraecum pyriforme	
Angraecum quintasii	**Diaphananthe rohrii**
Angraecum ramosum Thouars	
Angraecum ramosum auct. non Thouars, H.Perrier	**Angraecum germinyanum***
Angraecum ramosum ssp. *bidentatum*	**Angraecum germinyanum**
Angraecum ramosum ssp. *typicum*	**Angraecum germinyanum**
Angraecum ramosum ssp. *typicum* var. *arachnites*	**Angraecum germinyanum**
Angraecum ramosum ssp. *typicum* var. *bathiei*	**Angraecum germinyanum**
Angraecum ramosum ssp. *typicum* var. *conchoglossum*	**Angraecum germinyanum**
Angraecum ramosum ssp. *typicum* var. *peracuminatum*	**Angraecum germinyanum**
Angraecum ramulicolum	
Angraecum rectum	**Jumellea recta**

***For explanation see page 3, point 6**
***Voir les explications page 12, point 6**
***Para mayor explicación, véase la página 22, point 6**

ALL NAMES	ACCEPTED NAME
Angraecum recurvum	**Jumellea recurva**
Angraecum reichenbachianum	**Angraecum scottianum**
Angraecum reygaertii	
Angraecum rhipsalisocium	**Rangaeris rhipsalisocia**
Angraecum rhizanthium	
Angraecum rhizomaniacum	
Angraecum rhodesianum	**Tridactyle tricuspis**
Angraecum rhodostictum	**Aerangis luteo-alba** var. **rhodosticta**
Angraecum rhopaloceras	**Angraecum calceolus**
Angraecum rhynchoglossum	
Angraecum rigidifolium	
Angraecum ringens	**Cyrtorchis ringens**
Angraecum robustum Kraenzl. non Schltr.	**Angraecum kraenzlinianum**
Angraecum robustum Schltr.	**Sobennikoffia robusta***
Angraecum rohlfsianum	**Aerangis brachycarpa**
Angraecum rohrii	**Diaphananthe rohrii**
Angraecum roseocalcaratum	**Aerangis calantha**
Angraecum rostellare	**Aerangis rostellaris**
Angraecum rostratum	
Angraecum rothschildianum	**Eurychone rothschildiana**
Angraecum rubellum	
Angraecum rubrum	**Renanthera moluccana**
Angraecum rutenbergianum	
Angraecum sacciferum	
Angraecum saccolabioides	**Microterangis humblotii**
Angraecum sacculatum	
Angraecum salazianum	
Angraecum sambiranoense	
Angraecum sanderianum	**Aerangis modesta**
Angraecum sanfordii	
Angraecum sankuruense	**Aerangis calantha**
Angraecum sarcodanthum	**Angraecum crassum**
Angraecum saundersiae	**Aerangis mystacidii**
Angraecum scabripes	**Angraecum conchiferum**
Angraecum scalariforme	
Angraecum scandens	**Solenangis scandens**
Angraecum scandens ssp. *longifolia*	**Solenangis scandens**
Angraecum schiedei	**Campylocentrum schiedei**
Angraecum schimperianum Schweinf. non Rchb.f.	**Cyrtorchis erythraeae**
Angraecum schimperianum Rchb.f.	**Diaphananthe schimperiana***
Angraecum schoellerianum	**Calyptrochilum christyanum**
Angraecum schumannii	**Ancistrorhynchus schumannii**
Angraecum scottellii	**Tridactyle scottellii**
Angraecum scottianum	
Angraecum sedenii	**Cyrtorchis arcuata** ssp. **arcuata**
Angraecum sedifolium	
Angraecum sellowii	**Campylocentrum sellowii**
Angraecum semipedale	**Aerangis kotschyana**
Angraecum serpens	
Angraecum sesquipedale	
Angraecum sesquipedale var. **angustifolium**	
Angraecum sesquisectangulum	
Angraecum setipes	
Angraecum sinuatiflorum	
Angraecum smithii	**Microcoelia smithii**
Angraecum solheidii	**Rangaeris muscicola**
Angraecum somalense	**Aerangis somalensis**

***For explanation see page 3, point 6**
***Voir les explications page 12, point 6**
***Para mayor explicación, véase la página 22, point 6**

Part I: All Names / Tous les Noms / Todos los Nombres

ALL NAMES	ACCEPTED NAME
Angraecum sororium	
Angraecum spathulatum	**Jumellea spathulata**
Angraecum spectabile	
Angraecum spicatum	
Angraecum stella	**Aerangis gravenreuthii**
Angraecum stella-africae	
Angraecum stenophyllum	**Jumellea stenophylla**
Angraecum sterrophyllum	
Angraecum stipitatum	**Jumellea stipitata**
Angraecum stipulatum	**Tridactyle stipulata**
Angraecum stolzii	
Angraecum straussii	**Ancistrorhynchus straussii**
Angraecum striatum	
Angraecum stylosum	**Aerangis stylosa**
Angraecum suarezense	**Angraecum curnowianum**
Angraecum subclavatum	**Diaphananthe subclavata**
Angraecum subcordatum	**Angraecum curnowianum**
Angraecum subcylindrifolium	**Cyrtorchis aschersonii**
Angraecum subfalcifolium	**Diaphananthe bidens**
Angraecum subulatum	
Angraecum superbum	**Angraecum eburneum** ssp. **superbum**
Angraecum talbotii	**Bolusiella talbotii**
Angraecum tamarindicolum	
Angraecum tenellum	
Angraecum tenerrimum	**Angraecopsis tenerrima**
Angraecum tenue	**Campylocentrum tenue**
Angraecum tenuifolium	
Angraecum tenuipes	
Angraecum tenuispica	
Angraecum teres	
Angraecum teretifolium	
Angraecum thomense	**Chamaeangis thomensis**
Angraecum thomsonii	**Aerangis thomsonii**
Angraecum trachyrrhizum	**Tridactyle anthomaniaca**
Angraecum triangulifolium	
Angraecum trichoplectron	
Angraecum tricuspe	**Tridactyle tricuspis**
Angraecum tridactylites	**Tridactyle tridactylites**
Angraecum tridens	**Angraecopsis tridens**
Angraecum tridentatum	**Tridactyle tridentata**
Angraecum triquetrum	
Angraecum tsaratananae	**Angraecum zaratananae**
Angraecum umbrosum	
Angraecum undulatum	
Angraecum urostachyum	**Chamaeangis odoratissima**
Angraecum urschianum	
Angraecum vagans	**Chamaeangis vagans**
Angraecum venustum	**Aerangis articulata**
Angraecum verdickii	**Aerangis verdickii**
Angraecum verecundum	
Angraecum verrucosum	**Angraecum conchiferum**
Angraecum verruculosum sensu DuPuy	**Angraecum implicatum**
Angraecum verruculosum Frapp. ex Cordem.	**Angraecum ramosum***
Angraecum vesicatum	**Chamaeangis vesicata**
Angraecum vesiculatum	
Angraecum vesiculiferum	
Angraecum viguieri	

***For explanation see page 3, point 6**
***Voir les explications page 12, point 6**
***Para mayor explicación, véase la página 22, point 6**

ALL NAMES	ACCEPTED NAME
Angraecum virens	**Angraecum eburneum**
Angraecum virgula	**Tridactyle virgula**
Angraecum viride Kraenzl.	
Angraecum viride (Ridl.) Schltr.	**Angraecum rhynchoglossum***
Angraecum viridescens	**Tridactyle laurentii**
Angraecum viridiflorum	
Angraecum voeltzkowianum	**Angraecum eburneum** ssp. **superbum**
Angraecum wakefieldii	**Solenangis wakefieldii**
Angraecum waterlotii	**Angraecum tenellum**
Angraecum whitfieldii	**Tridactyle anthomaniaca**
Angraecum woodianum	**Diaphananthe rutila**
Angraecum xanthopollinium	**Diaphananthe xanthopollinia**
Angraecum xylopus	
Angraecum yuccaefolium	
Angraecum zaratananae	
Angraecum zenkeri	**Bolusiella zenkeri**
Angraecum zeylanicum	
Angraecum zigzag	**Calyptrochilum christyanum**
Anguloa lurida	**Catasetum luridum**
Anota densiflora	**Rhynchostylis gigantea**
Anota gigantea	**Rhynchostylis gigantea**
Anota hainanensis	**Rhynchostylis gigantea**
Anota harrisoniana	**Rhynchostylis gigantea**
Arachnis beccarii	**Vandopsis muelleri**
Arachnis muelleri	**Vandopsis muelleri**
Arethusa sinensis	**Bletilla sinensis**
Armodorum distichum	**Renanthera sulingi**
Armodorum sulingi	**Renanthera sulingi**
Ascocentrum ampullaceum	
Ascocentrum ampullaceum var. **auranticum**	
Ascocentrum aurantiacum	
Ascocentrum aurantiacum ssp. **philippinense**	
Ascocentrum aureum	
Ascocentrum christensonianum	
Ascocentrum curvifolium	
Ascocentrum garayi	
Ascocentrum hendersoniana	
Ascocentrum himalaicum	
Ascocentrum insularum	
Ascocentrum micranthum	**Smitinandia micrantha**
Ascocentrum miniatum	
Ascocentrum pumilum	
Ascocentrum pusillum	
Ascocentrum rubescens	**Aerides rubescens**
Ascocentrum rubrum	
Ascocentrum semiteretifolium	
Ascocentropsis pussilum	**Ascocentrum pusillum**
Ascolabium pumilum	**Ascocentrum pumilum**
Ascolabium pusillum	**Ascocentrum pusillum**
Aulostylis papuana	**Calanthe vestita**
Barombia gracillima	**Aerangis gracillima**
Bletia acaulis	**Brassavola acaulis**
Bletia amazonica	**Brassavola martiana**
Bletia angustata	**Brassavola angustata**
Bletia attenuata	**Brassavola angustata**
Bletia cebolleta	**Brassavola cebolleta**
Bletia cucullata	**Brassavola cucullata**

***For explanation see page 3, point 6**
***Voir les explications page 12, point 6**
***Para mayor explicación, véase la página 22, point 6**

ALL NAMES	ACCEPTED NAME
Bletia formosana	**Bletilla formosana**
Bletia gebina	**Bletilla striata**
Bletia hyacinthina	**Bletilla striata**
Bletia kotoensis	**Bletilla formosana**
Bletia lineata	**Brassavola acaulis**
Bletia masuca	**Calanthe sylvatica**
Bletia morrisonicola	**Bletilla formosana**
Bletia nodosa	**Brassavola nodosa**
Bletia perrinii	**Brassavola perrinii**
Bletia silvatica	**Calanthe sylvatica**
Bletia striata	**Bletilla striata**
Bletia venosa	**Brassavola venosa**
Bletilla burmanica	**Bletilla chartacea**
Bletilla chartacea	
Bletilla chinensis	**Bletilla sinensis**
Bletilla cotoensis	**Bletilla formosana**
Bletilla elegantula	**Bletilla formosana**
Bletilla florida	**Bletia florida**
Bletilla foliosa	
Bletilla formosana	
Bletilla formosana forma *kotoensis*	**Bletilla formosana**
Bletilla formosana forma *rubrolabella*	**Bletilla formosana**
Bletilla gebina	**Bletilla striata**
Bletilla hyacinthina	**Bletilla striata**
Bletilla japonica	**Eleorchis japonica**
Bletilla kotoensis	**Bletilla formosana**
Bletilla morrisonensis	**Bletilla formosana**
Bletilla morrisonicola	**Bletilla formosana**
Bletilla ochracea	
Bletilla scopulorum	**Pleione scopulorum**
Bletilla sinensis	
Bletilla striata (Thunb. ex A.Murray) Rchb.f.	
Bletilla striata (Thunb. ex A.Murray) Druce	**Bletilla striata**
Bletilla striata ssp. *albomarginata*	**Bletilla striata**
Bletilla striata forma *gebina*	**Bletilla striata**
Bletilla striata var. *kotoensis*	**Bletilla formosana**
Bletilla szetschuanica	**Bletilla formosana**
Bletilla yunnanensis	**Bletilla formosana**
Bletilla yunnanensis var. *limprichtii*	**Bletilla formosana**
Brassavola acaulis	
Brassavola amazonica	**Brassavola martiana**
Brassavola angustata	
Brassavola appendiculata	**Brassavola cucullata**
Brassavola cebolleta	
Brassavola cebolleta var. *fasciculata*	**Brassavola chacoensis**
Brassavola chacoensis	
Brassavola cordata	
Brassavola cucullata	
Brassavola cucullata var. *elegans*	**Brassavola cucullata**
Brassavola cuspidata	**Brassavola cucullata**
Brassavola digbyana	**Rhyncholaelia digbyana**
Brassavola duckeana	**Brassavola martiana**
Brassavola elegans	**Tetramicra canaliculata**
Brassavola elongata	**Brassavola cucullata**
Brassavola fasciculata	**Brassavola chacoensis**
Brassavola filifolia	
Brassavola flagellaris	

***For explanation see page 3, point 6**
***Voir les explications page 12, point 6**
***Para mayor explicación, véase la página 22, point 6**

ALL NAMES	ACCEPTED NAME
Brassavola fragans	**Brassavola gardneri**
Brassavola fragrans	**Brassavola tuberculata**
Brassavola gardneri	
Brassavola gibbsiana	**Brassavola tuberculata**
Brassavola gillettei	**Brassavola nodosa**
Brassavola glauca	**Rhyncholaelia glauca**
Brassavola grandiflora	
Brassavola harrisii	**Brassavola cordata**
Brassavola lineata	**Brassavola acaulis**
Brassavola martiana	
Brassavola martiana var. *multiflora*	**Brassavola martiana**
Brassavola mathieuana	**Brassavola acaulis**
Brassavola multiflora	**Brassavola martiana**
Brassavola nodosa (L.) Lindl.	
Brassavola nodosa Hook.	**Brassavola cordata**
Brassavola nodosa var. *cordata*	**Brassavola cordata**
Brassavola nodosa var. *grandiflora*	**Brassavola nodosa**
Brassavola nodosa var. *rhopalorrachis*	**Brassavola nodosa**
Brassavola nodosa var. *venosa*	**Brassavola venosa**
Brassavola odoratissima	**Brassavola cucullata**
Brassavola ovaliformis	**Brassavola chacoensis**
Brassavola ovaliformis var. *fasciculata*	**Brassavola chacoensis**
Brassavola paraensis	**Brassavola martiana**
Brassavola perrinii	
Brassavola perrinii var. *pluriflora*	**Brassavola perrinii**
Brassavola pescatorii	**Epidendrum parkinsonianum**
Brassavola pumilio	**Homalopetalum pumilio**
Brassavola reginae	**Brassavola cebolleta**
Brassavola retusa	
Brassavola revoluta Barb.Rodr.	**Brassavola tuberculata**
Brassavola revoluta sensu Withner	**Brassavola cebolleta**
Brassavola rhomboglossa	**Brassavola perrinii**
Brassavola rhopalorrachis	**Brassavola nodosa**
Brassavola rigida	**Tetramicra canaliculata**
Brassavola scaposa	**Brassavola nodosa**
Brassavola sloanei Griseb.	**Brassavola cordata**
Brassavola sloanei Lindl. ex Heynh.	**Brassavola nodosa***
Brassavola stricta	**Brassavola nodosa**
Brassavola suaveolens	**Epidendrum parkinsonianum**
Brassavola subulifolia	**Brassavola cordata**
Brassavola surinamensis	**Brassavola angustata**
Brassavola tuberculata	
Brassavola venosa	
Brassavola vomeriformis	**Homalopetalum vomeriforme**
Brassia clowesii	**Miltonia clowesii**
Calanthe abbreviata	
Calanthe aceras	
Calanthe actinomorpha	
Calanthe × albolilacina	
Calanthe albo-longicalcarata	
Calanthe albolutea	
Calanthe aleizettii	
Calanthe alismaefolia	
Calanthe alpina	
Calanthe alpina ssp. *fimbriata*	**Calanthe alpina**
Calanthe alpina ssp. *fimbriatomarginata*	**Calanthe alpina**
Calanthe alpina var. *keshabii*	**Calanthe alpina**

***For explanation see page 3, point 6**
***Voir les explications page 12, point 6**
***Para mayor explicación, véase la página 22, point 6**

Part I: All Names / Tous les Noms / Todos los Nombres

ALL NAMES	ACCEPTED NAME
Calanthe alta	
Calanthe amamiana	**Calanthe discolor** ssp. **amamiana**
Calanthe amamiana ssp. *latilabella*	**Calanthe discolor** ssp. **amamiana**
Calanthe amoena	**Calanthe puberula**
Calanthe anchorifera	**Calanthe triplicata**
Calanthe angraeciflora	**Calanthe triplicata**
Calanthe angusta	
Calanthe angusta	**Calanthe odora**
Calanthe angusta var. *laeta*	**Calanthe odora**
Calanthe angustifolia	
Calanthe anjanae	
Calanthe anocentrum	**Calanthe alta**
Calanthe anthropophora	
Calanthe apostasioides	**Calanthe caulescens**
Calanthe arcuata	
Calanthe arcuata var. **brevifolia**	
Calanthe arfakana	
Calanthe argenteostriata	
Calanthe arisanensis	
Calanthe aristulifera	
Calanthe aristulifera var. *amamiana*	**Calanthe discolor** ssp. **amamiana**
Calanthe aristulifera ssp. *izu-insularis*	**Calanthe izu-insularis**
Calanthe aruank	
Calanthe arundinoides	**Calanthe versteegii**
Calanthe atjehensis	
Calanthe augusti-reigneri	**Calanthe vestita**
Calanthe aurantiaca	
Calanthe aurantimacula	
Calanthe aureiflora	
Calanthe australis	**Spiranthes sinensis**
Calanthe austrokiusiuensis	**Calanthe alismaefolia**
Calanthe bachmaensis	**Phaius tankervilleae**
Calanthe balansae	
Calanthe baliensis	
Calanthe bicalcarata	
Calanthe bicolor	**Calanthe striata**
Calanthe bigibba	**Calanthe ventilabrum**
Calanthe biloba Lindl.	
Calanthe biloba auct. non Lindl.	**Calanthe cardioglossa***
Calanthe biloba ssp. *obtusa*	**Calanthe biloba**
Calanthe brachychila	
Calanthe bracteosa Rchb.f.	**Calanthe triplicata***
Calanthe bracteosa Schltr.	**Calanthe rhodochila**
Calanthe brevicolumna	**Calanthe triplicata**
Calanthe brevicornu	
Calanthe breviflos	
Calanthe breviscapa	**Calanthe rhodochila**
Calanthe buccinifera	
Calanthe bungoana	**Calanthe davidii**
Calanthe burmanica	**Calanthe ceciliae**
Calanthe bursicula	**Cephalantheropsis obcordata**
Calanthe calanthoides	
Calanthe camptoceras	
Calanthe candida	
Calanthe cardioglossa	
Calanthe carrii	
Calanthe catilligera	**Calanthe triplicata**

***For explanation see page 3, point 6**
***Voir les explications page 12, point 6**
***Para mayor explicación, véase la página 22, point 6**

ALL NAMES	ACCEPTED NAME
Calanthe caudatilabella	
Calanthe caudatilabella var. *latiloba*	**Calanthe caudatilabella**
Calanthe caulescens	
Calanthe caulodes	
Calanthe ceciliae	
Calanthe celebica	**Calanthe sylvatica**
Calanthe cheniana	**Calanthe discolor**
Calanthe chevalieri	
Calanthe chlorantha	**Collabium chloranthum**
Calanthe chloroleuca	
Calanthe chrysantha	**Calanthe ventilabrum**
Calanthe chrysoglossoides	
Calanthe chrysoleuca	
Calanthe citrina	**Calanthe striata**
Calanthe clavata	
Calanthe clavata var. *malipoensis*	**Calanthe clavata**
Calanthe clavicalcar	
Calanthe cleistogama	
Calanthe coelogyniformis	**Calanthe delavayi**
Calanthe coiloglossa	
Calanthe comosa	**Calanthe triplicata**
Calanthe conspicua	
Calanthe coreana	
Calanthe corymbosa	**Calanthe sylvatica**
Calanthe cremeoviridis	
Calanthe crenulata	
Calanthe crinita non Gagnep.	**Phaius indigoferus***
Calanthe crinita sensu Seidenf.	**Phaius mishmensis**
Calanthe cruciata	
Calanthe crumenata	
Calanthe cuneata	**Calanthe kinabaluensis**
Calanthe cubensis	**Calanthe calanthoides**
Calanthe curantigoides	**Calanthe pulchra**
Calanthe curculigoides	**Calanthe pulchra**
Calanthe curtisii	**Calanthe sylvatica**
Calanthe curvato-ascendens	
Calanthe davaensis	
Calanthe davidii	
Calanthe davidii var. *bungoana*	**Calanthe davidii**
Calanthe delavayi	
Calanthe delphinioides	**Calanthe sylvatica**
Calanthe densiflora Lindl.	
Calanthe densiflora sensu King & Pantling non Lindl.	**Calanthe clavata**
Calanthe diploxiphion	**Calanthe triplicata**
Calanthe dipteryx	
Calanthe discolor	
Calanthe discolor forma *bicolor*	**Calanthe striata**
Calanthe discolor forma **quinquelamellata**	
Calanthe discolor forma *viridialba*	**Calanthe discolor** ssp. **amamiana**
Calanthe discolor ssp. **amamiana**	
Calanthe discolor ssp. *bicolor*	**Calanthe striata**
Calanthe discolor ssp. **discolor**	
Calanthe discolor ssp. *divaricatipetala*	**Calanthe discolor** ssp. **discolor**
Calanthe discolor ssp. *flava*	**Calanthe striata**
Calanthe discolor ssp. **kanashiroi**	
Calanthe discolor ssp. *latilabella*	**Calanthe discolor** ssp. **amamiana**
Calanthe discolor ssp. **tokunoshimensis**	

***For explanation see page 3, point 6**
***Voir les explications page 12, point 6**
***Para mayor explicación, véase la página 22, point 6**

Part I: All Names / Tous les Noms / Todos los Nombres

ALL NAMES	ACCEPTED NAME
Calanthe discolor ssp. *viridialba*	**Calanthe discolor** ssp. **discolor**
Calanthe discolor var. *amamiana*	**Calanthe discolor** ssp. **amamiana**
Calanthe disticha	**Calanthe formosana**
Calanthe dolichopoda	**Cephalantheropsis longipes**
Calanthe dulongensis	
Calanthe durani	**Calanthe sylvatica**
Calanthe eberhardtii	
Calanthe ecallosa	
Calanthe ecarinata	
Calanthe elliptica	**Calanthe aristulifera**
Calanthe elmeri	**Calanthe rubens**
Calanthe elytroglossa	**Calanthe triplicata**
Calanthe emarginata	**Calanthe sylvatica**
Calanthe emeishanica	
Calanthe engleriana	
Calanthe engleriana var. *brevicalcarata*	**Calanthe engleriana**
Calanthe englishii	**Calanthe ventilabrum**
Calanthe ensifolia	**Calanthe davidii**
Calanthe epiphytica	
Calanthe esquirolii	**Calanthe discolor**
Calanthe fargesii	
Calanthe fauriei	**Calanthe alismaefolia**
Calanthe fimbriata	**Calanthe alpina**
Calanthe fimbriatomarginata	**Calanthe alpina**
Calanthe finisterrae	
Calanthe fissa	
Calanthe flabelliformis	**Calanthe ventilabrum**
Calanthe flava	
Calanthe foerstermannii	**Calanthe lyroglossa**
Calanthe forbesii	
Calanthe formosana	
Calanthe forsythiiflora	**Calanthe lyroglossa**
Calanthe fragrans	**Calanthe micrantha**
Calanthe fragilis	**Calanthe tenuis**
Calanthe fuerstenbergiana	**Calanthe cardioglossa**
Calanthe fulgens	
Calanthe furcata	**Calanthe triplicata**
Calanthe furcata var. *alismifolia*	**Calanthe alismaefolia**
Calanthe furcata forma *albo-lineata*	**Calanthe triplicata**
Calanthe furcata forma *albo-marginata*	**Calanthe triplicata**
Calanthe furcata forma *brevicolumna*	**Calanthe triplicata**
Calanthe furcata forma *faurie*	**Calanthe alismaefolia**
Calanthe furcata forma *masuca*	**Calanthe sylvatica**
Calanthe furcata forma *matsudai*	**Calanthe davidii**
Calanthe furcata forma *raishaensis*	**Calanthe aristulifera**
Calanthe furcata forma *textorii*	**Calanthe sylvatica**
Calanthe galeata	**Calanthe chloroleuca**
Calanthe gebina	**Bletilla striata**
Calanthe geelvinkensis	
Calanthe gibbsiae	
Calanthe gigantea	**Plocoglottis gigantea**
Calanthe gougahensis	**Acanthephippium gougahense**
Calanthe graciliflora	
Calanthe graciliflora var. *xuefengensis*	**Calanthe graciliflora**
Calanthe gracilis	**Cephalantheropsis obcordata**
Calanthe gracilis var. *venusta*	**Cephalantheropsis obcordata**
Calanthe graciliscapa	

***For explanation see page 3, point 6**
***Voir les explications page 12, point 6**
***Para mayor explicación, véase la página 22, point 6**

ALL NAMES	ACCEPTED NAME
Calanthe gracillima	**Calanthe triplicata**
Calanthe granatensis	**Calanthe calanthoides**
Calanthe grandiflora	**Calanthe vestita**
Calanthe griffithii	
Calanthe halconensis	
Calanthe hamata	**Calanthe graciliflora**
Calanthe hancockii	
Calanthe hattorii	
Calanthe hennisii	
Calanthe henryi	
Calanthe herbacea	
Calanthe hirsuta	
Calanthe hololeuca	
Calanthe hoshii	
Calanthe hosseusiana	**Calanthe cardioglossa**
Calanthe humbertii	
Calanthe hyacinthina	
Calanthe imthurnii	
Calanthe inflata	
Calanthe integrilabris	**Calanthe velutina**
Calanthe inquilinax	**Calanthe vestita**
Calanthe izu-insularis	
Calanthe japonica	**Calanthe alismaefolia**
Calanthe johorensis	
Calanthe jusnerii	
Calanthe kaniensis	
Calanthe kawakamii	**Calanthe striata**
Calanthe kazuoi	**Calanthe densiflora**
Calanthe kemulense	
Calanthe keshabii	**Calanthe alpina**
Calanthe kinabaluensis	
Calanthe kintaroi	**Calanthe sylvatica**
Calanthe kirishimensis	**Calanthe aristulifera**
Calanthe kooshunensis	**Cephalantheropsis halconensis**
Calanthe labellicauda	
Calanthe labrosa	
Calanthe lacerata	
Calanthe lamellata	**Calanthe tricarinata**
Calanthe lamellosa	**Calanthe brevicornu**
Calanthe langei	**Calanthe ventilabrum**
Calanthe latissimifolia	**Calanthe engleriana**
Calanthe laxiflora	
Calanthe lechangensis	
Calanthe lepida	**Calanthe puberula**
Calanthe leucosceptrum	
Calanthe lilacina	**Calanthe conspicua**
Calanthe limprichtii	
Calanthe liukiuensis	**Calanthe lyroglossa**
Calanthe longibracteata	
Calanthe longicalcarata	**Calanthe sylvatica**
Calanthe longifolia	
Calanthe longipes	**Cephalantheropsis longipes**
Calanthe lurida	**Calanthe discolor** ssp. **discolor**
Calanthe lutescens	**Calanthe alta**
Calanthe lutiviridis	
Calanthe lyroglossa	
Calanthe lyroglossa var. *forsythiiflora*	**Calanthe lyroglossa**

***For explanation see page 3, point 6**
***Voir les explications page 12, point 6**
***Para mayor explicación, véase la página 22, point 6**

ALL NAMES	ACCEPTED NAME
Calanthe lyroglossa var. *longibracteata*	**Calanthe lyroglossa**
Calanthe madagascariensis	
Calanthe manis	**Calanthe rhodochila** var. **reconditiflora**
Calanthe mannii	
Calanthe maquilingensis	
Calanthe masuca	**Calanthe sylvatica**
Calanthe masuca forma *albiflora*	**Calanthe sylvatica**
Calanthe masuca var. *fulgens*	**Calanthe fulgens**
Calanthe masuca var. *purpurea*	**Calanthe sylvatica**
Calanthe masuca var. *sinensis*	**Calanthe sylvatica**
Calanthe matsumurana	**Calanthe triplicata**
Calanthe matsudai	**Calanthe davidii**
Calanthe mcgregorii	
Calanthe megalopha	**Calanthe tricarinata**
Calanthe melinosema	
Calanthe metoensis	
Calanthe mexicana	**Calanthe calanthoides**
Calanthe mexicana var. *retusa*	**Calanthe calanthoides**
Calanthe mexicana var. *lanceolata*	**Calanthe calanthoides**
Calanthe micrantha	
Calanthe microglossa	
Calanthe millotae	
Calanthe mindorensis	
Calanthe moluccensis	
Calanthe monophylla	
Calanthe muelleri	**Calanthe triplicata**
Calanthe mutabilis	**Calanthe zollingeri**
Calanthe nankunensis	
Calanthe natalensis	**Calanthe sylvatica**
Calanthe neglecta	**Calanthe sylvatica**
Calanthe neocaledonica	**Calanthe hololeuca**
Calanthe neohibernica	**Calanthe hololeuca**
Calanthe nephroglossa	
Calanthe nephroidea	**Calanthe lyroglossa**
Calanthe nicolae	
Calanthe nigropuncticulata	**Calanthe alismaefolia**
Calanthe nipponica	
Calanthe nivalis	
Calanthe oblanceolata	**Calanthe mannii**
Calanthe obreniformis	
Calanthe occidentalis	**Calanthe tricarinata**
Calanthe odora	
Calanthe okinawensis	**Calanthe sylvatica**
Calanthe okushirensis	**Calanthe reflexa**
Calanthe ombrophila	**Calanthe micrantha**
Calanthe × oodaruma	
Calanthe × oreadum	
Calanthe orthocentron	**Calanthe triplicata**
Calanthe otuhanica	
Calanthe ovalifolia	
Calanthe ovata	
Calanthe pachystalix Rchb.f.	
Calanthe pachystalix Gagnep.	**Calanthe angusta**
Calanthe padangensis	**Calanthe vestita**
Calanthe pantlingii	**Calanthe tricarinata**
Calanthe papuana	**Calanthe vestita**
Calanthe parviflora	**Calanthe flava**

ALL NAMES	ACCEPTED NAME
Calanthe parvilabris	
Calanthe patsinensis	**Calanthe formosana**
Calanthe pauciverrucosa	
Calanthe pavairiensis	
Calanthe perrottetii	**Calanthe triplicata**
Calanthe perrieri	**Calanthe sylvatica**
Calanthe petelotiana	
Calanthe petri	**Calanthe triplicata**
Calanthe phajoides	**Calanthe angustifolia**
Calanthe pilosa	**Calanthe vestita**
Calanthe plantaginea* Lindl.	
Calanthe plantaginea Griff.	**Calanthe griffithii**
Calanthe plantaginea var. *lushuiensis*	**Calanthe plantaginea**
Calanthe pleiochroma	**Calanthe sylvatica**
Calanthe poilanei	
Calanthe polyantha	
Calanthe proboscidea	**Calanthe triplicata**
Calanthe puberula	
Calanthe puberula var. *reflexa*	**Calanthe reflexa**
Calanthe puberula var. *caudatilabella*	**Calanthe caudatilabella**
Calanthe puberula var. *formosana*	**Calanthe formosana**
Calanthe puberula var. *okushirensis*	**Calanthe reflexa**
Calanthe pubescens	
Calanthe pulchra	
Calanthe pulchra var. *formosana*	**Calanthe formosana**
Calanthe pullei	
Calanthe pumila	
Calanthe purpurea	**Calanthe sylvatica**
Calanthe pusilla Carr.	**Calanthe carrii**
Calanthe pusilla Finet.	**Calanthe mannii***
Calanthe raishaensis	**Calanthe aristulifera**
Calanthe rajana	
Calanthe ramosa	**Phaius mishmensis**
Calanthe ramosii	**Cephalantheropsis obcordata**
Calanthe reconditiflora	**Calanthe rhodochila** var. **reconditiflora**
Calanthe reflexa	
Calanthe reflexa ssp. *okushirensis*	**Calanthe reflexa**
Calanthe reflexilabris	
Calanthe regnieri	**Calanthe vestita**
Calanthe repens	
Calanthe repens ssp. *pauliani*	**Calanthe repens**
Calanthe rhodochila	
Calanthe rhodochila var. **reconditiflora**	
Calanthe rigida	
Calanthe rosea	
Calanthe rubens	
Calanthe rubicallosa	**Calanthe triplicata**
Calanthe ruttenii	
Calanthe saccata	
Calanthe saccifera	**Calanthe balansae**
Calanthe sacculata	
Calanthe sacculata var. *tchenkeoutinensis*	**Calanthe sacculata**
Calanthe salaccensis	
Calanthe salmoniviridis	**Calanthe caulescens**
Calanthe sanderiana	**Calanthe sylvatica**
Calanthe sasakii	**Calanthe arisanensis**
Calanthe scaposa	**Calanthe brevicornu**

***For explanation see page 3, point 6**
***Voir les explications page 12, point 6**
***Para mayor explicación, véase la página 22, point 6**

ALL NAMES	ACCEPTED NAME
Calanthe schlechteri	**Calanthe alpina**
Calanthe schliebenii	**Calanthe sylvatica**
Calanthe scortechini	**Calanthe ceciliae**
Calanthe scortechinii	**Calanthe lyroglossa**
Calanthe seikooensis	**Calanthe sylvatica**
Calanthe seranica	
Calanthe shelfordii	
Calanthe shweliensis	**Calanthe odora**
Calanthe sieboldii	**Calanthe striata**
Calanthe similis	**Calanthe reflexa**
Calanthe simplex	
Calanthe sinica	
Calanthe spathoglottoides	**Calanthe sylvatica**
Calanthe speciosa (Blume) Lindl.	
Calanthe speciosa Viell.	**Calanthe speciosa**
Calanthe stenophylla	
Calanthe stevensiana	**Calanthe vestita**
Calanthe stevensii	**Calanthe vestita**
Calanthe stolzii	**Calanthe sylvatica**
Calanthe striata R.Br. ex Lindl	
Calanthe striata Lindl.	**Calanthe striata**
Calanthe striata ssp. *bicolor*	**Calanthe striata**
Calanthe striata ssp. *sieboldii*	**Calanthe striata**
Calanthe striata var. *pumila*	**Calanthe pumila**
Calanthe × subhamata	
Calanthe succedanea	
Calanthe sumatrana	**Calanthe ceciliae**
Calanthe sylvatica* (Thouars) Lindl.	
Calanthe sylvatica sensu Rolfe	**Calanthe madagascariensis**
Calanthe sylvatica ssp. *natalensis*	**Calanthe sylvatica**
Calanthe sylvatica var. *pallidipetala*	**Calanthe sylvatica**
Calanthe sylvestris	**Calanthe sylvatica**
Calanthe taenioides	
Calanthe tahitensis	
Calanthe tangmaiensis	
Calanthe takeoi	**Calanthe striata**
Calanthe tenuis	
Calanthe textori	**Calanthe sylvatica**
Calanthe textori ssp. *alba*	**Calanthe sylvatica**
Calanthe textori forma *albiflora*	**Calanthe sylvatica**
Calanthe textori ssp. *longicalcarata*	**Calanthe sylvatica**
Calanthe textori ssp. *textori*	**Calanthe sylvatica**
Calanthe textori ssp. *violacea*	**Calanthe sylvatica**
Calanthe tokunoshimensis	**Calanthe discolor** ssp. **tokunoshimensis**
Calanthe tokunoshimensis forma *latilabella*	**Calanthe discolor** ssp. **tokunoshimensis**
Calanthe torifera	**Calanthe tricarinata**
Calanthe torricellensis	
Calanthe transiens	
Calanthe triantherifera	
Calanthe tricarinata	
Calanthe trifida	
Calanthe triplicata	
Calanthe triplicata forma *albo-lineata*	**Calanthe triplicata**
Calanthe triplicata forma *albo-marginata*	**Calanthe triplicata**
Calanthe triplicata var. *angraeciflora*	**Calanthe triplicata**
Calanthe triplicata var. *gracillima*	**Calanthe triplicata**
Calanthe trulliformis ssp. *hastata*	**Calanthe nipponica**

***For explanation see page 3, point 6**
***Voir les explications page 12, point 6**
***Para mayor explicación, véase la página 22, point 6**

ALL NAMES	ACCEPTED NAME
Calanthe truncata	
Calanthe truncicola	
Calanthe tsoongiana	
Calanthe tsoongiana var. *guizhouensis*	**Calanthe tsoongiana**
Calanthe tubifera	**Cephalantheropsis obcordata**
Calanthe tunensis	**Calanthe ventilabrum**
Calanthe turneri	**Calanthe vestita**
Calanthe tyoh-harai	
Calanthe uncata	
Calanthe undulata Schltr.	**Calanthe tricarinata***
Calanthe undulata J.J.Sm.	
Calanthe unifolia	
Calanthe vaginata	**Calanthe odora**
Calanthe × varians	
Calanthe variegata	**Calanthe discolor** ssp. **discolor**
Calanthe vaupeliana	**Calanthe hololeuca**
Calanthe velutina	
Calanthe ventilabrum	
Calanthe venusta	**Cephalantheropsis obcordata**
Calanthe veratrifolia (Willd.) Ker Gawl.	**Calanthe triplicata***
Calanthe veratrifolia Miq. non R.Br.	**Calanthe flava**
Calanthe veratrifolia ssp. *densissima*	**Calanthe triplicata**
Calanthe veratrifolia ssp. *dupliciloba*	**Calanthe triplicata**
Calanthe veratrifolia ssp. *incurvicalca*	**Calanthe triplicata**
Calanthe veratrifolia ssp. *incurvicalcar*	**Calanthe triplicata**
Calanthe veratrifolia ssp. *lancipetala*	**Calanthe triplicata**
Calanthe veratrifolia ssp. *timorensis*	**Calanthe triplicata**
Calanthe veratrifolia var. *australis*	**Calanthe triplicata**
Calanthe veratrifolia var. *kennyi*	**Calanthe triplicata**
Calanthe veratrifolia var. *stenochila*	**Calanthe triplicata**
Calanthe versteegii	
Calanthe versicolor	**Calanthe sylvatica**
Calanthe vestita	
Calanthe vestita ssp. *fournieri*	**Calanthe vestita**
Calanthe vestita var. *fournieri* sensu Seidenf.	**Calanthe rubens**
Calanthe vestita ssp. *igneo-oculata*	**Calanthe vestita**
Calanthe vestita ssp. *oculata-gigantea*	**Calanthe vestita**
Calanthe villosa	
Calanthe viridifusca	**Ania viridifusca**
Calanthe violacea	**Calanthe sylvatica**
Calanthe volkensii	**Calanthe sylvatica**
Calanthe wardii	**Calanthe whiteana**
Calanthe warpuri	**Calanthe madagascariensis**
Calanthe whiteana	
Calanthe wightii	**Calanthe sylvatica**
Calanthe wrayi	**Calanthe ceciliae**
Calanthe yoksomnensis	
Calanthe yuana	
Calanthe yunnanensis	**Calanthe brevicornu**
Calanthe yushunii	**Calanthe formosana**
Calanthe zollingeri Rchb.f.	
Calanthe zollingeri Miq. (non Rchb.f.)	**Calanthe ceciliae***
Calanthe zollingeri var. *longecalcarata*	**Calanthe zollingeri**
Calanthidium labrosum	**Calanthe labrosa**
Catachaetum craniomorphum	**Catasetum luridum**
Catachaetum lituratum	**Catasetum luridum**
Catachaetum purum	**Catasetum purum**

***For explanation see page 3, point 6**
***Voir les explications page 12, point 6**
***Para mayor explicación, véase la página 22, point 6**

Part I: All Names / Tous les Noms / Todos los Nombres

ALL NAMES	ACCEPTED NAME
Catachaetum recurvatum	**Catasetum planiceps**
Catachaetum purpurascens	**Catasetum luridum**
Catachaetum semiapertum	**Catasetum purum**
Catachaetum squalidum	**Catasetum luridum**
Catachaetum turbinatum	**Catasetum luridum**
Catasetum abruptum	**Catasetum luridum**
Catasetum acallosum	**Catasetum callosum**
Catasetum aculeatum	
Catasetum adnatum	**Catasetum atratum**
Catasetum adremedium	
Catasetum albopurpureum	**Catasetum × tapiriceps**
Catasetum albovirens	
Catasetum apertum	**Catasetum × tapiriceps**
Catasetum appendiculatum Schltr. sensu Dunst.	**Catasetum barbatum***
Catasetum appendiculatum Schltr. sensu K.G.Lacerda	**Catasetum lanciferum**
Catasetum appendiculatum	**Catasetum gladiatorium**
Catasetum arachnoides	**Catasetum callosum**
Catasetum aripuanense	
Catasetum atratum	
Catasetum atratum var. *mentosum*	**Catasetum atratum**
Catasetum baraquinianum	**Catasetum saccatum**
Catasetum barbatum	
Catasetum barbatum var. *probiscidium*	**Catasetum barbatum**
Catasetum barbatum var. *spinosum*	**Catasetum barbatum**
Catasetum bergoldianum	
Catasetum bicallosum	
Catasetum bicolor	
Catasetum bicornutum	**Catasetum cornutum**
Catasetum blackii	
Catasetum blepharochilum	**Catasetum maculatum**
Catasetum boyi	
Catasetum brachybulbon Schltr.	
Catasetum brachybulbon Schltr. sensu Dunst.	**Catasetum barbatum***
Catasetum brenesii	**Catasetum maculatum**
Catasetum brichtiae	**Catasetum ferox**
Catasetum buchtienii Kraenzl.	
Catasetum buchtienii Kraenzl. sensu Dunst.	**Catasetum barbatum**
Catasetum bungerothi	**Catasetum pileatum**
Catasetum bungerothi var. *album*	**Catasetum pileatum**
Catasetum bungerothi var. *aurantiacum*	**Catasetum pileatum**
Catasetum bungerothi var. *lindeni*	**Catasetum pileatum**
Catasetum bungerothi var. *pottsianum*	**Catasetum pileatum**
Catasetum bungerothi var. *randi*	**Catasetum pileatum**
Catasetum bungerothi var. *regale*	**Catasetum pileatum**
Catasetum bungerothii var. *imperiale*	**Catasetum pileatum**
Catasetum cabrutae	**Catasetum × tapiriceps**
Catasetum calceolatum	**Clowesia russelliana**
Catasetum callosum	
Catasetum callosum var. *crenatum*	**Catasetum callosum**
Catasetum callosum var. *carunculatum*	**Catasetum callosum**
Catasetum callosum var. *eucallosum*	**Catasetum callosum**
Catasetum callosum var. *grandiflorum*	**Catasetum callosum**
Catasetum callosum var. *typum*	**Catasetum callosum**
Catasetum caputinum	
Catasetum carolinianum	
Catasetum carunculatum	**Catasetum callosum**

***For explanation see page 3, point 6**
***Voir les explications page 12, point 6**
***Para mayor explicación, véase la página 22, point 6**

ALL NAMES	ACCEPTED NAME
Catasetum cassideum	**Catasetum discolor**
Catasetum caucanum	**Catasetum tabulare**
Catasetum cernuum	
Catasetum cernuum var. *cernuum*	**Catasetum cernuum**
Catasetum cernuum var. *revolutum*	**Catasetum cernuum**
Catasetum cernuum var. *rodigasianum*	**Catasetum cernuum**
Catasetum cernuum var. *typum*	**Catasetum cernuum**
Catasetum cernuum var. *umbrosum*	**Catasetum cernuum**
Catasetum charlesworthii	
Catasetum chloranthum	**Catasetum × sodiroi**
Catasetum christyanum	**Catasetum saccatum**
Catasetum christyanum var. *chlorops*	**Catasetum saccatum**
Catasetum christyanum var. *obscurum*	**Catasetum saccatum**
Catasetum ciliatum	**Catasetum × roseo-album**
Catasetum cirrhaeoides	**Catasetum pulchrum**
Catasetum cirrhaeoides var. *hoehnei*	**Catasetum pulchrum**
Catasetum cirrhaeoides var. *longicirrhosa*	**Catasetum pulchrum**
Catasetum claesianum	**Catasetum discolor**
Catasetum claveringii	**Catasetum macrocarpum**
Catasetum cliftonii	**Catasetum expansum**
Catasetum cochabambanum	
Catasetum cogniauxii	**Catasetum fimbriatum**
Catasetum collare	
Catasetum colossus	**Catasetum saccatum**
Catasetum comminatum	nomen nudum
Catasetum comosum	**Catasetum barbatum**
Catasetum complanatum	
Catasetum confusum	
Catasetum coniforme	
Catasetum cornutum	
Catasetum costatum	
Catasetum cotylicheilum	
Catasetum crinitum	**Catasetum barbatum**
Catasetum craniomorphum	**Catasetum barbatum**
Catasetum crashleyanum	**Catasetum barbatum**
Catasetum cristatum	
Catasetum cristatum var. *monstruosum*	**Catasetum cristatum**
Catasetum cristatum var. *spinigerum*	**Catasetum barbatum**
Catasetum cristatum var. *spinosum*	**Catasetum barbatum**
Catasetum cristatum var. *stenosepalum*	**Catasetum cristatum**
Catasetum cristatum var. *supralobatum*	**Catasetum cristatum**
Catasetum cruciatum	**Catasetum saccatum**
Catasetum cucullatum	
Catasetum cupuliforme	**Catasetum × tapiriceps**
Catasetum darwinianum	**Catasetum callosum**
Catasetum decipiens	
Catasetum deltoideum	
Catasetum denticulatum	
Catasetum dilectum	**Dressleria dilecta**
Catasetum dilectum var. *suave*	**Dressleria dilecta**
Catasetum discolor	
Catasetum discolor forma *genuinum*	**Catasetum discolor**
Catasetum discolor var. *bushnani*	**Catasetum discolor**
Catasetum discolor var. *claesianum*	**Catasetum discolor**
Catasetum discolor var. *discolor*	**Catasetum discolor**
Catasetum discolor var. *roseo-album*	**Catasetum × roseo-album**

***For explanation see page 3, point 6**
***Voir les explications page 12, point 6**
***Para mayor explicación, véase la página 22, point 6**

Part I: All Names / Tous les Noms / Todos los Nombres

ALL NAMES	ACCEPTED NAME
Catasetum discolor var. *vinosum*	**Catasetum × roseo-album**
Catasetum discolor var. *viridiflorum*	**Catasetum discolor**
Catasetum dodsonianum	**Clowesia dodsoniana**
Catasetum × dunstervillei	
Catasetum duplicisculatum	
Catasetum eburneum	**Dressleria eburnea**
Catasetum expansum	
Catasetum faustii	
Catasetum fernandezii	
Catasetum ferox	
Catasetum fimbriatum* (Morren) Lindl. & Paxton	
Catasetum fimbriatum Rchb.f. non Lindl.	**Catasetum × roseo-album**
Catasetum fimbriatum var. *aurantiacum*	**Catasetum fimbriatum**
Catasetum fimbriatum var. *brevipetalum*	**Catasetum fimbriatum**
Catasetum fimbriatum var. *cogniauxii*	**Catasetum fimbriatum**
Catasetum fimbriatum var. *fissum*	**Catasetum fimbriatum**
Catasetum fimbriatum var. *inconstans*	**Catasetum fimbriatum**
Catasetum fimbriatum var. *micranthum*	**Catasetum fimbriatum**
Catasetum fimbriatum var. *morrenianum*	**Catasetum fimbriatum**
Catasetum fimbriatum var. *ornithorrhynchum*	**Catasetum fimbriatum**
Catasetum fimbriatum var. *platypterum*	**Catasetum fimbriatum**
Catasetum fimbriatum var. *subtropicale*	**Catasetum fimbriatum**
Catasetum fimbriatum var. *viridulum*	**Catasetum fimbriatum**
Catasetum finetianum	**Catasetum tabulare**
Catasetum floribundum	**Catasetum macrocarpum**
Catasetum franchinianum	
Catasetum fuchsii	
Catasetum fuliginosum	**Catasetum callosum**
Catasetum galeatum	
Catasetum galeritum	
Catasetum galeritum var. *galeritum*	**Catasetum galeritum**
Catasetum galeritum var. *pachyglossum*	**Catasetum galeritum**
Catasetum gardneri	**Catasetum discolor**
Catasetum garnettianum Rolfe	
Catasetum garnettianum Rolfe sensu Dunst.	**Catasetum barbatum**
Catasetum georgii	
Catasetum gladiatorium	
Catasetum glaucoglossum	**Clowesia glaucoglossa**
Catasetum globiferum	**Catasetum globiflorum**
Catasetum globiflorum	
Catasetum gnomus	
Catasetum gnomus var. *phasma*	**Catasetum gnomus**
Catasetum gomezii	
Catasetum gongoroides	**Catasetum bicolor**
Catasetum grandis	**Catasetum macrocarpum**
Catasetum guentherianum	**Mormodes guentheriana**
Catasetum × guianense	
Catasetum heteranthum	**Catasetum gnomus**
Catasetum hillsii	
Catasetum histrio	**Catasetum saccatum**
Catasetum hoehnei	**Catasetum tigrinum**
Catasetum hookeri	
Catasetum hookeri var. *labiatum*	**Catasetum hookeri**
Catasetum hookeri var. *triste*	**Catasetum hookeri**
Catasetum huebneri Schltr.	
Catasetum huebneri Mansf. non Schltr.	**Catasetum georgii**

***For explanation see page 3, point 6**
***Voir les explications page 12, point 6**
***Para mayor explicación, véase la página 22, point 6**

ALL NAMES	ACCEPTED NAME
Catasetum huebneri sensu Romero & Jenny	**Catasetum gnomus***
Catasetum hymenophorum	**Catasetum planiceps**
Catasetum immaculatum	**Catasetum purum**
Catasetum imperiale	
Catasetum imschootianum	**Catasetum hookeri**
Catasetum inapertum	**Catasetum purum**
Catasetum inconstans	**Catasetum fimbriatum**
Catasetum incurvum	
Catasetum inornatum	**Catasetum ochraceum**
Catasetum integerrimum	
Catasetum integerrimum var. *purpurascens*	**Catasetum integerrimum**
Catasetum integerrimum var. *viridiflorum*	**Catasetum integerrimum**
Catasetum integerrimum var. *flavescens*	**Catasetum macrocarpum**
Catasetum integerrimum var. *luteo-purpureum*	**Catasetum × tapiriceps**
Catasetum integrinum	**Catasetum integerrimum**
Catasetum × intermedium	
Catasetum intermedium var. *rubrum*	**Catasetum × intermedium**
Catasetum intermedium var. *zebrinum*	**Catasetum × intermedium**
Catasetum × issanensis	
Catasetum japurense	**Catasetum saccatum**
Catasetum jarae	**Catasetum pulchrum**
Catasetum juruense	
Catasetum justinianum	
Catasetum kempfii	
Catasetum kleberianum	
Catasetum kraenzlinianum	
Catasetum labiatum	**Catasetum hookeri**
Catasetum lamilatum	**Catasetum laminatum**
Catasetum laminatum	
Catasetum laminatum var. *eburneum*	**Catasetum laminatum**
Catasetum laminatum var. *maculatum*	**Catasetum laminatum**
Catasetum lanceanum	
Catasetum lanciferum	
Catasetum landsbergii	**Catasetum callosum**
Catasetum lanxiforme	
Catasetum lehmannii	**Catasetum ochraceum**
Catasetum lemosii	
Catasetum liechtensteinii	**Catasetum trulla**
Catasetum lindenii	**Catasetum × tapiriceps**
Catasetum lindleyanum	
Catasetum linguiferum	
Catasetum lituratum	**Catasetum luridum**
Catasetum longifolium	
Catasetum longipes	
Catasetum lucianii	**Catasetum × tapiriceps**
Catasetum lucis	
Catasetum lucis forma *tigrinum*	**Catasetum lucis**
Catasetum luridum	
Catasetum macrocarpum Rich. ex Kunth.	
Catasetum macrocarpum Stein non Rich. ex Kunth.	**Catasetum barbatum**
Catasetum macrocarpum var. *amplissimum*	**Catasetum macrocarpum**
Catasetum macrocarpum var. *aurantiacum*	**Catasetum × tapiriceps**
Catasetum macrocarpum var. *bellum*	**Catasetum macrocarpum**
Catasetum macrocarpum var. *brevifolium*	**Catasetum macrocarpum**
Catasetum macrocarpum var. *carnosissimum*	**Catasetum macrocarpum**
Catasetum macrocarpum var. *chrysanthum* Lind & Rod.	**Catasetum macrocarpum**

***For explanation see page 3, point 6**
***Voir les explications page 12, point 6**
***Para mayor explicación, véase la página 22, point 6**

Part I: All Names / Tous les Noms / Todos los Nombres

ALL NAMES	ACCEPTED NAME
Catasetum macrocarpum var. *chrysanthum* Hort ex W.H.G.	**Catasetum × tapiriceps**
Catasetum macrocarpum var. *flavescens*	**Catasetum × tapiriceps**
Catasetum macrocarpum var. *genuinum*	**Catasetum macrocarpum**
Catasetum macrocarpum var. *globoso-connivens*	**Catasetum macrocarpum**
Catasetum macrocarpum var. *lindeni*	**Catasetum macrocarpum**
Catasetum macrocarpum var. *luteo-purpureum*	**Catasetum × tapiriceps**
Catasetum macrocarpum var. *luteo-roseum*	**Catasetum × tapiriceps**
Catasetum macrocarpum var. *pallidum*	**Catasetum macrocarpum**
Catasetum macrocarpum var. *unidentatum*	**Catasetum macrocarpum**
Catasetum macrocarpum var. *viridi-eburneum*	**Catasetum macrocarpum**
Catasetum macrocarpum var. *viridi-sanguineum*	**Catasetum macrocarpum**
Catasetum macroglossum	
Catasetum maculatum* Kunth	
Catasetum maculatum auct. non Kunth	**Catasetum integerrimum**
Catasetum maculatum var. *flavescens*	**Catasetum macrocarpum**
Catasetum maculatum var. *luteo-purpureum*	**Catasetum × tapiriceps**
Catasetum magnificum	**Catasetum × tapiriceps**
Catasetum maranhense	
Catasetum maroaense	
Catasetum mattogrossense	
Catasetum mattosianum	
Catasetum meeae	
Catasetum menthaeodorum	**Catasetum macrocarpum**
Catasetum mentosum	**Catasetum atratum**
Catasetum merchae	
Catasetum micranthum Barb.Rodr.	
Catasetum micranthum Kraenzl. non Barb.Rodr.	**Catasetum kraenzlinianum**
Catasetum microglossum	
Catasetum milleri	**Catasetum hookeri**
Catasetum mirabile	**Catasetum × tapiriceps**
Catasetum mocuranum	
Catasetum mojuense	
Catasetum monodon	**Catasetum triodon**
Catasetum monzonensis	
Catasetum moorei	
Catasetum multifidum	
Catasetum multifissum	
Catasetum nanayanum	
Catasetum napoense	
Catasetum naso Lindl.	
Catasetum naso Hook non Lindl.	**Catasetum sanguineum***
Catasetum naso var. *charlesworthii*	**Catasetum charlesworthii**
Catasetum naso var. *naso*	**Catasetum naso**
Catasetum naso var. *pictum*	**Catasetum sanguineum**
Catasetum naso var. *viride*	**Catasetum sanguineum**
Catasetum negrense	**Catasetum fimbriatum**
Catasetum o'brienianum	**Catasetum × tapiriceps**
Catasetum ochraceum	
Catasetum oerstedii	**Catasetum maculatum**
Catasetum ollare	
Catasetum ornithoides	
Catasetum ornithorhynchum	**Catasetum fimbriatum**
Catasetum osculatum	
Catasetum pallidiflorum	**Catasetum tabulare**
Catasetum pallidum	**Catasetum atratum**

***For explanation see page 3, point 6**
***Voir les explications page 12, point 6**
***Para mayor explicación, véase la página 22, point 6**

ALL NAMES	ACCEPTED NAME
Catasetum parguazense	
Catasetum pendulum	
Catasetum peruvianum	
Catasetum pflanzii	**Catasetum fimbriatum**
Catasetum phasma	**Catasetum gnomus**
Catasetum pileatum	
Catasetum pileatum var. *album*	**Catasetum pileatum**
Catasetum pileatum var. *aurantiacum*	**Catasetum pileatum**
Catasetum pileatum var. *imperiale*	**Catasetum pileatum**
Catasetum pileatum var. *lindeni*	**Catasetum pileatum**
Catasetum pileatum var. *pottisianum*	**Catasetum pileatum**
Catasetum pileatum var. *randii*	**Catasetum pileatum**
Catasetum pileatum var. *regale*	**Catasetum pileatum**
Catasetum planiceps	
Catasetum platyglossum	**Catasetum expansum**
Catasetum platyglossum var. *sodiroi*	**Catasetum × sodiroi**
Catasetum pleiodactylon	
Catasetum × pohlianum	
Catasetum polydactylon Schltr.	
Catasetum polydactylon Schltr. sensu Dunst.	**Catasetum barbatum***
Catasetum poriferum	
Catasetum proboscideum	**Catasetum barbatum**
Catasetum pulchrum	
Catasetum punctatum	
Catasetum purpurascens	**Catasetum luridum**
Catasetum purum	
Catasetum purusense	
Catasetum pusillum C.Schweinf.	
Catasetum pusillum C.Schweinf. fide Brako, L. & J.L.Zarucchi	**Catasetum × rose-album**
Catasetum quadricolor	**Catasetum × tapiriceps**
Catasetum quadridens	
Catasetum quornus	nomen nudum
Catasetum randii Rolfe	
Catasetum randii Rolfe sensu Dunst.	**Catasetum barbatum***
Catasetum recurvatum	**Catasetum planiceps**
Catasetum regnellii	
Catasetum reichenbachianum	
Catasetum revolutum	**Catasetum × tapiriceps**
Catasetum rhamphastos	**Catasetum tabulare**
Catasetum richteri	
Catasetum ricii	
Catasetum rivularium Barb.Rodr.	
Catasetum rivularium Barb.Rodr. sensu Dunst.	**Catasetun barbatum***
Catasetum rodigasianum	**Catasetum cernuum**
Catasetum rohrii	**Catasetum cernuum**
Catasetum rolfeanum	
Catasetum rondonense	
Catasetum rooseveltianum	
Catasetum × roseo-album	
Catasetum roseum Barb.Rodr. non (Lindl.) Rchb.f.	**Catasetum lemosii**
Catasetum roseum (Lindl.) Rchb.f.	**Clowesia rosea***
Catasetum rostratum	**Catasetum maculatum**
Catasetum russellianum	**Clowesia russelliana**
Catasetum saccatum	
Catasetum saccatum var. *album*	**Catasetum saccatum**

***For explanation see page 3, point 6**
***Voir les explications page 12, point 6**
***Para mayor explicación, véase la página 22, point 6**

Part I: All Names / Tous les Noms / Todos los Nombres

ALL NAMES	ACCEPTED NAME
Catasetum saccatum var. *chlorops*	**Catasetum saccatum**
Catasetum saccatum var. *christyanum*	**Catasetum saccatum**
Catasetum saccatum var. *eusaccatum*	**Catasetum saccatum**
Catasetum saccatum var. *incurvum*	**Catasetum saccatum**
Catasetum saccatum var. *pliciferum*	**Catasetum saccatum**
Catasetum saccatum var. *typum*	**Catasetum saccatum**
Catasetum saccatum var. *viride*	**Catasetum saccatum**
Catasetum samaniegoi	
Catasetum sanguineum	
Catasetum sanguineum var. *integrale*	**Catasetum sanguineum**
Catasetum sanguineum var. *viride*	**Catasetum sanguineum**
Catasetum schmidtianum	
Catasetum schunkei	
Catasetum schweinfurthii	
Catasetum scurra	**Clowesia warscewiczii**
Catasetum semiapertum	**Catasetum purum**
Catasetum semicirculatum	
Catasetum semiroseum	**Catasetum × tapiriceps**
Catasetum serratum	**Catasetum viridiflavum**
Catasetum socco	**Catasetum trulla**
Catasetum × sodiroi	
Catasetum spinosum (Hook.) Lindl.	
Catasetum spinosum Lindl. sensu Dunst.	**Catasetum barbatum***
Catasetum spitzii	
Catasetum spitzii var. *album*	**Catasetum spitzii**
Catasetum spitzii var. *roseum*	**Catasetum spitzii**
Catasetum spitzii var. *sanguineum*	**Catasetum spitzii**
Catasetum splendens	**Catasetum × tapiriceps**
Catasetum splendens var. *acutipetalum*	**Catasetum × tapiriceps**
Catasetum splendens var. *albo-purpureum*	**Catasetum × tapiriceps**
Catasetum splendens var. *album*	**Catasetum × tapiriceps**
Catasetum splendens var. *aliciae*	**Catasetum × tapiriceps**
Catasetum splendens var. *atropurpureum*	**Catasetum × tapiriceps**
Catasetum splendens var. *aurantiacum*	**Catasetum × tapiriceps**
Catasetum splendens var. *aureo-maculatum*	**Catasetum × tapiriceps**
Catasetum splendens var. *aureum*	**Catasetum × tapiriceps**
Catasetum splendens var. *eburneum*	**Catasetum × tapiriceps**
Catasetum splendens var. *flavescens*	**Catasetum × tapiriceps**
Catasetum splendens var. *griganii*	**Catasetum × tapiriceps**
Catasetum splendens var. *imperiale*	**Catasetum × tapiriceps**
Catasetum splendens var. *lansbergianum*	**Catasetum × tapiriceps**
Catasetum splendens var. *lindeni*	**Catasetum × tapiriceps**
Catasetum splendens var. *luciani*	**Catasetum × tapiriceps**
Catasetum splendens var. *maculatum*	**Catasetum × tapiriceps**
Catasetum splendens var. *mirabile*	**Catasetum × tapiriceps**
Catasetum splendens var. *o'brienianum*	**Catasetum × tapiriceps**
Catasetum splendens var. *punctatissimum*	**Catasetum × tapiriceps**
Catasetum splendens var. *regale*	**Catasetum × tapiriceps**
Catasetum splendens var. *revolutum*	**Catasetum × tapiriceps**
Catasetum splendens var. *rubrum*	**Catasetum × tapiriceps**
Catasetum splendens var. *semiroseum*	**Catasetum × tapiriceps**
Catasetum splendens var. *viride*	**Catasetum × tapiriceps**
Catasetum splendens var. *worthingtonianum*	**Catasetum × tapiriceps**

***For explanation see page 3, point 6**
***Voir les explications page 12, point 6**
***Para mayor explicación, véase la página 22, point 6**

ALL NAMES	ACCEPTED NAME
Catasetum stenochilum	**Catasetum rolfeanum**
Catasetum stenoglossum	**Catasetum bicallosum**
Catasetum stevensonii	
Catasetum stupendum	**Catasetum incurvum**
Catasetum suave	**Dressleria suavis**
Catasetum tabulare	
Catasetum tabulare var. *brachyglossum*	**Catasetum tabulare**
Catasetum tabulare var. *finetianum*	**Catasetum tabulare**
Catasetum tabulare var. *laeve*	**Catasetum tabulare**
Catasetum tabulare var. *pallidum*	**Catasetum tabulare**
Catasetum tabulare var. *rhamphastos*	**Catasetum tabulare**
Catasetum tabulare var. *rhinophorum*	**Catasetum tabulare**
Catasetum tabulare var. *rugosum*	**Catasetum tabulare**
Catasetum tabulare var. *serrulata*	**Catasetum tabulare**
Catasetum tabulare var. *virens*	**Catasetum tabulare**
Catasetum taguariense	
Catasetum taguariense var. *album*	**Catasetum taguariense**
Catasetum × tapiriceps	
Catasetum taquariense	
Catasetum tenebrosum	
Catasetum tenebrosum forma *smaragdinum*	**Catasetum tenebrosum**
Catasetum tenuiglossum	
Catasetum thompsonii	
Catasetum thylaciochilum	**Clowesia thylaciochila**
Catasetum tigrinum	
Catasetum transversicallosum	
Catasetum trautmannii	**Catasetum incurvum**
Catasetum tricolor Hort. ex Planchon	**Catasetum macrocarpum**
Catasetum tricolor Rchb.f. (spham)	**Catasetum triodon**
Catasetum tricorne	
Catasetum tridentatum Hook.	**Catasetum macrocarpum**
Catasetum tridentatum Pfitzer non Hook.	**Catasetum barbatum**
Catasetum tridentatum var. *amplissimum*	**Catasetum macrocarpum**
Catasetum tridentatum var. *breviflorum*	**Catasetum macrocarpum**
Catasetum tridentatum var. *claveringii*	**Catasetum macrocarpum**
Catasetum tridentatum var. *floribundrum*	**Catasetum macrocarpum**
Catasetum tridentatum var. *geniunum*	**Catasetum macrocarpum**
Catasetum tridentatum var. *globoso-connivens*	**Catasetum macrocarpum**
Catasetum tridentatum var. *macrocarpum*	**Catasetum macrocarpum**
Catasetum tridentatum var. *maximum*	**Catasetum macrocarpum**
Catasetum tridentatum var. *pallidum*	**Catasetum macrocarpum**
Catasetum tridentatum var. *unidentatum*	**Catasetum macrocarpum**
Catasetum tridentatum var. *viridi-eburnem*	**Catasetum macrocarpum**
Catasetum tridentatum var. *viridiflorum*	**Catasetum macrocarpum**
Catasetum tridentatum var. *viridi-sanguinem*	**Catasetum macrocarpum**
Catasetum trifidum	**Catasetum cernuum**
Catasetum trilobatum	**Catasetum × sodiroi**
Catasetum trimerochilum	**Mormodes lineata**
Catasetum triodon	
Catasetum triodon var. *guttulatum*	**Catasetum triodon**
Catasetum triste	**Catasetum hookeri**
Catasetum trulla	
Catasetum trulla var. *liechtensteinii*	**Catasetum trulla**
Catasetum trulla var. *maculatissimum*	**Catasetum trulla**
Catasetum trulla var. *subimberbe*	**Catasetum trulla**
Catasetum trulla var. *trilobatum*	**Catasetum trulla**

***For explanation see page 3, point 6**
***Voir les explications page 12, point 6**
***Para mayor explicación, véase la página 22, point 6**

Part I: All Names / Tous les Noms / Todos los Nombres

ALL NAMES	ACCEPTED NAME
Catasetum trulla var. *trulla*	**Catasetum trulla**
Catasetum trulla var. *typum*	**Catasetum trulla**
Catasetum trulla var. *vinaceum*	**Catasetum vinaceum**
Catasetum tuberculatum	
Catasetum turbinatum	
Catasetum tururuiense	
Catasetum umbrosum	**Catasetum cernuum**
Catasetum uncatum	
Catasetum vibratile	
Catasetum variabile	**Catasetum barbatum**
Catasetum vinaceum	
Catasetum vinaceum var. *album*	**Catasetum vinaceum**
Catasetum vinaceum var. *splendidum*	**Catasetum vinaceum**
Catasetum violascens	
Catasetum viride	**Catasetum cernuum**
Catasetum viridiflavum	
Catasetum wailesii	**Catasetum integerrimum**
Catasetum warscewiczii	**Clowesia warscewiczii**
Catasetum × wendlingeri	
Catasetum wredeanum	**Catasetum fimbriatum**
Catasetum yavitaense	
Centrosia auberti	**Calanthe sylvatica**
Centrosis sylvatica	**Calanthe sylvatica**
Cephalanthera chartacea	**Bletilla chartacea**
Coelogyne elegantula	**Bletilla formosana**
Ctenorchis pectinata	**Angraecum pectinatum**
Cymbidium allagnata	**Vanda spathulata**
Cymbidium cucullatum	**Brassavola cucullata**
Cymbidium furvum	**Vanda concolor**
Cymbidium hyacinthinum	**Bletilla striata**
Cymbidium nodosum	**Brassavola nodosa**
Cymbidium spathulatum	**Vanda spathulata**
Cymbidium striatum	**Bletilla striata**
Cymbidium tesselatum	**Vanda tessellata**
Cymbidium tesseloides	**Vanda tessellata**
Cypripedium cothurnum	**Catasetum trulla**
Cypripedium socco	**Catasetum trulla**
Cyrtochilum flavescens	**Miltonia flavescens**
Cyrtochilum karwinskii	**Miltonioides karwinskii**
Cyrtochilum stellatum	**Miltonia flavescens**
Cytheris griffithii	**Calanthe vestita**
Dendrobium brachycarpum	**Aerangis brachycarpa**
Dyakia hendersoniana	**Ascocentrum hendersoniana**
Eggelingia ligulifolia	**Angraecum marii**
Epidendrum coriaceum	**Angraecum coriaceum**
Epidendrum cucullatum	**Brassavola cucullata**
Epidendrum furvum	**Vanda concolor**
Epidendrum hippium	**Rhynchostylis retusa**
Epidendrum indicum	**Rhynchostylis retusa**
Epidendrum nodosum	**Brassavola nodosa**
Epidendrum ollare	**Catasetum luridum**
Epidendrum renanthera	**Renanthera coccinea**
Epidendrum retusum	**Rhynchostylis retusa**
Epidendrum spathulatum	**Vanda spathulata**
Epidendrum striatum	**Bletilla striata**
Epidendrum tesselatum	**Vanda tessellata**

***For explanation see page 3, point 6**
***Voir les explications page 12, point 6**
***Para mayor explicación, véase la página 22, point 6**

ALL NAMES	ACCEPTED NAME
Epidendrum tesseloides	**Vanda spathulata**
Epidendrum tuberosum	**Bletilla striata**
Epidorchis calceolus	**Angraecum calceolus**
Epidorchis caulescens	**Angraecum caulescens**
Epidorchis graminifolia	**Angraecum pauciramosum**
Epidorchis inaperta	**Angraecum inapertum**
Epidorchis multiflora	**Angraecum multiflorum**
Epidorchis subulata	**Angraecum subulatum**
Epidorchis tenella	**Angraecum tenellum**
Epidorchis viridis	**Angraecum rhynchoglossum**
Esmerelda sanderiana	**Vanda sanderiana**
Esmerelda sanderiana var. *albata*	**Vanda sanderiana**
Esmerelda sanderiana var. *labello-viridi*	**Vanda sanderiana**
Euanthe sanderiana	**Vanda sanderiana**
Fieldia gigantea	**Vandopsis gigantea**
Fieldia lissochiloides	**Vandopsis lissochiloides**
Fieldia undulata	**Vandopsis undulata**
Flos triplicatus	**Calanthe triplicata**
Gastrochilus ampullaceus	**Ascocentrum ampullaceum**
Gastrochilus blumei	**Rhynchostylis retusa**
Gastrochilus coriaceus	**Angraecum coriaceum**
Gastrochilus curvifolius	**Ascocentrum curvifolium**
Gastrochilus gurwalicus	**Rhynchostylis retusa**
Gastrochilus miniatus	**Ascocentrum miniatum**
Gastrochilus praemorsus	**Rhynchostylis retusa**
Gastrochilus retusus	**Rhynchostylis retusa**
Gastrochilus rheedii	**Rhynchostylis retusa**
Gastrochilus spicatus	**Rhynchostylis retusa**
Gastrochilus strictus	**Angraecum striatum**
Ghiesbreghtia calanthoides	**Calanthe calanthoides**
Ghiesbreghtia mexicana	**Calanthe calanthoides**
Gongora philippica	**Renanthera coccinea**
Grammatophyllum pantherinum	**Vandopsis lissochiloides**
Gyas humilis	**Bletilla striata**
Holcoglossum junceum	**Ascocentrum himalaicum**
Jimensia formosana	**Bletilla formosana**
Jimensia kotoensis	**Bletilla formosana**
Jimensia morrisonenensis	**Bletilla formosana**
Jimensia nervosa	**Bletilla striata**
Jimensia ochracea	**Bletilla ochracea**
Jimensia sinensis	**Bletilla sinensis**
Jimensia striata	**Bletilla striata**
Jimensia yunnanensis	**Bletilla formosana**
Jumellea curnowiana	**Angraecum curnowianum**
Jumellea humberti	**Angraecum ampullaceum**
Jumellea meirax	**Angraecum meirax**
Jumellea rutenbergiana	**Angraecum rutenbergianum**
Jumellea serpens	**Angraecum serpens**
Jumellea subcordata	**Angraecum curnowianum**
Lepervenchea tenuifolia	**Angraecum tenuifolium**
Leptocentrum spiculatum	**Aerangis spiculata**
Limatodes labrosa	**Calanthe labrosa**
Limatodis rosea	**Calanthe rosea**
Limodorum coriaceum	**Angraecum coriaceum**
Limodorum eburneum	**Angraecum eburneum**
Limodorum retusa	**Rhynchostylis retusa**
Limodorum spathulatum	**Vanda spathulata**

Part I: All Names / Tous les Noms / Todos los Nombres

ALL NAMES	ACCEPTED NAME
Limodorum striatum Banks (non Thunb.)	**Calanthe striata**
Limodorum striatum Thunb.	**Bletilla striata***
Limodorum striatum Reinw. ex Hook.f.	**Calanthe angustifolia**
Limodorum ventricosum	**Calanthe triplicata**
Limodorum veratrifolium	**Calanthe triplicata**
Listrostachys biloba	**Aerangis biloba**
Listrostachys bracteosa	**Angraecum bracteosum**
Listrostachys clavata	**Angraecum multinominatum**
Listrostachys palmiformis	**Angraecum palmiforme**
Listrostachys subulata	**Angraecum subulatum**
Luisia alpina	**Vanda alpina**
Lysimmia bicolor	**Brassavola cordata**
Macrochilus fryanus	**Miltonia spectabilis**
Macroplectrum baronii	**Angraecum baronii**
Macroplectrum calceolus	**Angraecum calceolus**
Macroplectrum clavatum	**Angraecum multinominatum**
Macroplectrum costatum	**Angraecum costatum**
Macroplectrum cucullatum	**Angraecum cucullatum**
Macroplectrum didieri	**Angraecum didieri**
Macroplectrum distichophyllum	**Angraecum striatum**
Macroplectrum gladiifolium	**Angraecum mauritianum**
Macroplectrum humblotii	**Angraecum humblotianum**
Macroplectrum implicatum	**Angraecum implicatum**
Macroplectrum leonis	**Angraecum leonis**
Macroplectrum madagascariense	**Angraecum madagascariense**
Macroplectrum meirax	**Angraecum meirax**
Macroplectrum ochraceum	**Angraecum ochraceum**
Macroplectrum pectinatum	**Angraecum pectinatum**
Macroplectrum sesquipedale	**Angraecum sesquipedale**
Macroplectrum xylopus	**Angraecum xylopus**
Miltonia anceps	**Miltonia flava**
Miltonia bicolor	**Miltonia spectabilis**
Miltonia × binotii	
Miltonia bismarckii	**Miltoniopsis bismarckii**
Miltonia × bluntii	
Miltonia candida	**Anneliesia candida**
Miltonia candida var. *flavescens*	**Anneliesia candida**
Miltonia candida var. *purpureo-violacea*	**Anneliesia candida**
Miltonia × castanea	
Miltonia cereola	**Miltonia regnellii**
Miltonia clowesii	
Miltonia clowesii var. *castanea*	**Miltonia castanea**
Miltonia clowesii var. *lamarckeana*	**Miltonia × lamarckeana**
Miltonia × cogniauxiae	
Miltonia cogniauxiae var. *bicolor*	**Miltonia × cogniauxiae**
Miltonia cogniauxiae var. *massaiana*	**Miltonia × cogniauxiae**
Miltonia cogniauxiae var. *pallida*	**Miltonia × cogniauxiae**
Miltonia cuneata	**Anneliesia cuneata**
Miltonia cyrtochiloides	**Miltonia × festiva**
Miltonia endresii	**Miltoniopsis warscewiczii**
Miltonia × festiva	
Miltonia flava	
Miltonia flavescens	
Miltonia flavescens var. *grandiflora*	**Miltonia flavescens**
Miltonia flavescens var. *stellata*	**Miltonia flavescens**

***For explanation see page 3, point 6**
***Voir les explications page 12, point 6**
***Para mayor explicación, véase la página 22, point 6**

ALL NAMES	ACCEPTED NAME
Miltonia flavescens var. *typica*	**Miltonia flavescens**
Miltonia joiceyana	**Miltonia × lamarckeana**
Miltonia karwinskii	**Miltonioides karwinskii**
Miltonia kayasimae	**Anneliesia kayasimae**
Miltonia laevis sensu Brieger & Lükel	**Miltonioides laevis***
Miltonia laevis sensu Rolfe	**Odontoglossum laeve**
Miltonia × lamarckeana	
Miltonia lawrenceana	**Miltonia × castanea**
Miltonia × leucoglossa	
Miltonia leucomelas	**Miltonioides leucomelas**
Miltonia loddigesii	**Miltonia flavescens**
Miltonia moreliana	**Miltonia spectabilis**
Miltonia odorata	
Miltonia parva	**Cischweinfia parva**
Miltonia peetersiana	**Miltonia × bluntii**
Miltonia phalaenopsis	**Miltoniopsis phalaenopsis**
Miltonia pulchella	**Miltoniopsis phalaenopsis**
Miltonia pinelli	**Miltonia flava**
Miltonia quadrijuga	**Anneliesia russeliana**
Miltonia regnellii	
Miltonia regnelli var. *travassosiana*	**Miltonia regnellii**
Miltonia regnelli var. *veitchiana*	**Miltonia regnellii**
Miltonia reichenheimii	**Miltonioides reichenheimii**
Miltonia roezlii	**Miltoniopsis roezlii**
Miltonia roezlii ssp. *alba*	**Miltoniopsis roezlii** ssp. **alba**
Miltonia roezlii subvar. *alba*	**Miltoniopsis roezlii** ssp. **alba**
Miltonia rosea	**Miltonia spectabilis**
Miltonia × rosina	
Miltonia russelliana	**Anneliesia russeliana**
Miltonia schroederiana	**Miltonioides schroederiana**
Miltonia schroederianum	**Odontoglossum confusum**
Miltonia speciosa	**Anneliesia cuneata**
Miltonia spectabilis	
Miltonia spectabilis var. *lineata*	**Miltonia spectabilis**
Miltonia spectabilis var. *morelliana*	**Miltonia spectabilis**
Miltonia spectabilis var. *virginalis*	**Miltonia spectabilis**
Miltonia stellata	**Miltonia flavescens**
Miltonia stenoglossa sensu Brieger & Lükel	**Miltonioides leucomelas**
Miltonia stenoglossa sensu L.O.Williams	**Odontoglossum stenoglossum**
Miltonia superba	**Miltoniopsis warscewiczii**
Miltonia velloziana	**Anneliesia cuneata**
Miltonia vexillaria	**Miltoniopsis vexillaria**
Miltonia vexillaria var. *alba*	**Miltoniopsis vexillaria**
Miltonia vexillaria var. *carolina*	**Miltoniopsis vexillaria**
Miltonia vexillaria var. *hilliana*	**Miltoniopsis vexillaria**
Miltonia vexillaria var. *kienastiana*	**Miltoniopsis vexillaria**
Miltonia vexillaria var. *leopoldii*	**Miltoniopsis vexillaria**
Miltonia vexillaria var. *leucoglossa*	**Miltoniopsis vexillaria**
Miltonia vexillaria var. *stupenda*	**Miltoniopsis vexillaria**
Miltonia vexillaria var. *superba*	**Miltoniopsis vexillaria**
Miltonia vexillaria var. *rosea*	**Miltoniopsis vexillaria**
Miltonia vexillaria var. *rubella*	**Miltoniopsis vexillaria**
Miltonia warneri	**Miltonia spectabilis**
Miltonia warscewiczii	**Chamaelorchis warscewiczii**
Miltonioides carinifera	
Miltonioides confusa	**Miltonioides schroederiana**

***For explanation see page 3, point 6**
***Voir les explications page 12, point 6**
***Para mayor explicación, véase la página 22, point 6**

Part I: All Names / Tous les Noms / Todos los Nombres

ALL NAMES	ACCEPTED NAME
Miltonioides karwinskii	
Miltonioides laevis	
Miltonioides leucomelas	
Miltonioides oviedomotae	
Miltonioides pauciflora	**Miltonioides leucomelas**
Miltonioides reichenheimii	
Miltonioides schroederiana	
Miltonioides stenoglossum	**Miltonioides leucomelas**
Miltonioides warscewiczii	**Chamaelorchis warscewiczii**
Miltoniopsis bismarckii	
Miltoniopsis phalaenopsis	
Miltoniopsis roezlii	
Miltoniopsis roezlii ssp. **alba**	
Miltoniopsis santanaei Garay & Dunst.	
Miltoniopsis santanaei sensu Jøgensen	**Miltoniopsis roezlii***
Miltoniopsis vexillaria	
Miltoniopsis warscewiczii	
Monachanthus bushnani	**Catasetum discolor**
Monachanthus cristatus	**Catasetum cristatum**
Monachanthus discolor	**Catasetum discolor**
Monachanthus discolor var. *bushnani*	**Catasetum discolor**
Monachanthus discolor var. *viridiflorus*	**Catasetum discolor**
Monachanthus fimbriatus	**Catasetum × roseo-album**
Monachanthus longifolius	**Catasetum longifolium**
Monachanthus roseo-albus	**Catasetum × roseo-album**
Monachanthus viridis	**Catasetum cernuum**
Monixus aporum	**Angraecum podochiloides**
Monixus clavigera	**Angraecum clavigerum**
Monixus graminifolius	**Angraecum pauciramosum**
Monixus multiflorus	**Angraecum multiflorum**
Monixus striatus	**Angraecum striatum**
Monixus teretifolius	**Angraecum teretifolium**
Myanthus barbatus	**Catasetum barbatum**
Myanthus callosus	**Catasetum callosum**
Myanthus cernuus	**Catasetum cernuum**
Myanthus cristatus	**Catasetum cristatum**
Myanthus deltoideus	**Catasetum deltoideum**
Myanthus fimbriatus	**Catasetum fimbriatum**
Myanthus grandiflorum	**Catasetum callosum**
Myanthus landsbergii	**Catasetum callosum**
Myanthus sanguineus	**Catasetum sanguineum**
Myanthus spinosus Hook.	**Catasetum spinosum**
Myanthus spinosus Hook. sensu Dunst.	**Catasetum barbatum***
Myanthus warscewiczii	**Clowesia warscewiczii**
Mystacidium angustum	**Angraecum angustum**
Mystacidium appendiculatum	**Aerangis appendiculata**
Mystacidium arthrophyllum	**Angraecum pungens**
Mystacidium astroarche	**Angraecum astroarche**
Mystacidium calceolus	**Angraecum calceolus**
Mystacidium caulescens	**Angraecum caulescens**
Mystacidium caulescens ssp. *multiflorum*	**Angraecum multiflorum**
Mystacidium cilaosianum	**Angraecum cilaosianum**
Mystacidium clavatum	**Angraecum multinominatum**
Mystacidium costatum	**Angraecum costatum**
Mystacidium crassifolium	**Angraecum crassifolium**
Mystacidium curnowianus	**Angraecum curnowianum**

***For explanation see page 3, point 6**
***Voir les explications page 12, point 6**
***Para mayor explicación, véase la página 22, point 6**

ALL NAMES	ACCEPTED NAME
Mystacidium dauphinense	**Angraecum dauphinense**
Mystacidium distichum	**Angraecum distichum**
Mystacidium distichum ssp. *grandifolium*	**Angraecum aporoides**
Mystacidium germinyanum	**Angraecum germinyanum**
Mystacidium gladiifolium	**Angraecum mauritianum**
Mystacidium graminifolium	**Angraecum pauciramosum**
Mystacidium gravenreuthii	**Aerangis gravenreuthii**
Mystacidium hermanni	**Angraecum hermannii**
Mystacidium inapertum	**Angraecum inapertum**
Mystacidium infundibulare	**Angraecum infundibulare**
Mystacidium keniae	**Angraecum keniae**
Mystacidium leonis	**Angraecum leonis**
Mystacidium longinode	**Angraecum longinode**
Mystacidium mauritianum	**Angraecum mauritianum**
Mystacidium minutum	**Angraecum minutum**
Mystacidium multiflorum	**Angraecum multiflorum**
Mystacidium nanum	**Angraecum nanum**
Mystacidium obversifolium	**Angraecum obversifolium**
Mystacidium ochraceum	**Angraecum ochraceum**
Mystacidium pectinatum	**Angraecum pectinatum**
Mystacidium pingue	**Angraecum pingue**
Mystacidium pseudo-petiolatum	**Angraecum pseudopetiolatum**
Mystacidium salazianum	**Angraecum salazianum**
Mystacidium sesquipedale	**Angraecum sesquipedale**
Mystacidium spicatum	**Angraecum spicatum**
Mystacidium striatum	**Angraecum cordemoyi**
Mystacidium tenellum	**Angraecum tenellum**
Mystacidium trichoplectron	**Angraecum trichoplectron**
Mystacidium undulatum	**Angraecum undulatum**
Mystacidium verrucosum	**Angraecum conchiferum**
Mystacidium viride	**Angraecum rhynchoglossum**
Nephranthera matutina	**Renanthera matutina**
Odontoglossum anceps	**Miltonia flava**
Odontoglossum cariniferum	**Miltonioides carinifera**
Odontoglossum clowesii	**Miltonia clowesii**
Odontoglossum confusum	**Miltonioides schroederiana**
Odontoglossum confusum var. *album*	**Miltonioides schroederiana**
Odontoglossum grande	**Rossioglossum grande**
Odontoglossum grande var. *aureum*	**Rossioglossum grande** var. **aureum**
Odontoglossum grande var. *citrinum*	**Rossioglossum grande** var. **aureum**
Odontoglossum grande var. *excelsior*	**Rossioglossum grande**
Odontoglossum grande var. *flavidum*	**Rossioglossum schlieperianum** var. **flavidum**
Odontoglossum grande var. *hibernum*	**Rossioglossum grande**
Odontoglossum grande var. *magnificum*	**Rossioglossum grande**
Odontoglossum grande var. *pallidum*	**Rossioglossum schlieperianum** var. **flavidum**
Odontoglossum grande var. *pittianum*	**Rossioglossum grande** var. **aureum**
Odontoglossum grande var. *sanderae*	**Rossioglossum grande** var. **aureum**
Odontoglossum grande var. *splendens*	**Rossioglossum grande**
Odontoglossum grande var. *superbum*	**Rossioglossum grande**
Odontoglossum grande var. *williamsianum*	**Rossioglossum williamsianum**
Odontoglossum insleayi	**Rossioglossum insleayi**
Odontoglossum insleayi var. *aureum*	**Rossioglossum splendens** var. **imschootianum**
Odontoglossum insleayi var. *imschootianum*	**Rossioglossum splendens** var. **imschootianum**
Odontoglossum insleayi var. *leopardinum*	**Rossioglossum splendens** var. **leopardinum**
Odontoglossum insleayi var. *macranthum*	**Rossioglossum schlieperianum**
Odontoglossum insleayi var. *pantherinum*	**Rossioglossum splendens** var. **pantherinum**
Odontoglossum insleayi var. *splendens*	**Rossioglossum splendens**

***For explanation see page 3, point 6**
***Voir les explications page 12, point 6**
***Para mayor explicación, véase la página 22, point 6**

ALL NAMES	ACCEPTED NAME
Odontoglossum karwinskii	**Miltonioides karwinskii**
Odontoglossum laeve	**Miltonioides laevis**
Odontoglossum laeve var. *auratum*	**Miltonioides leucomelas**
Odontoglossum laeve var. *karwinskii*	**Miltonioides karwinskii**
Odontoglossum laeve var. *reichenheimii*	**Miltonioides reichenheimii**
Odontoglossum lawrenceanum Hort. ex Gard.	**Rossioglossum insleayi**
Odontoglossum lawrenceanum Hort.	**Rossioglossum schlieperianum**
Odontoglossum leucomelas	**Miltonioides leucomelas**
Odontoglossum pauciflorum	**Miltonioides leucomelas**
Odontoglossum phalaenopsis	**Miltoniopsis phalaenopsis**
Odontoglossum powellii	**Rossioglossum powellii**
Odontoglossum reichenheimii	**Miltonioides reichenheimii**
Odontoglossum roezlii	**Miltoniopsis roezlii**
Odontoglossum roezlii var. *album*	**Miltoniopsis roezlii** ssp. **alba**
Odontoglossum schlieperianum	**Rossioglossum schlieperianum**
Odontoglossum schlieperianum var. *actuinum*	**Rossioglossum schlieperianum** var. **flavidum**
Odontoglossum schlieperianum var. *citrinum*	**Rossioglossum schlieperianum** var. **flavidum**
Odontoglossum schlieperianum var. *flavidum*	**Rossioglossum schlieperianum** var. **flavidum**
Odontoglossum schlieperianum var. *pretiosum*	**Rossioglossum powellii**
Odontoglossum schlieperianum var. *xauthinum*	**Rossioglossum schlieperianum** var. **flavidum**
Odontoglossum schroederianum	**Miltonioides schroederiana**
Odontoglossum splendens	**Rossioglossum splendens**
Odontoglossum stenoglossum	**Miltonioides leucomelas**
Odontoglossum vexillarium	**Miltoniopsis vexillaria**
Odontoglossum warscewiczianum	**Miltoniopsis warscewiczii**
Odontoglossum warscewiczii (Rchb.f.) Garay & Dunst.	**Miltoniopsis warscewiczii**
Odontoglossum warscewiczii sensu Bridges	**Rossioglossum schlieperianum***
Odontoglossum williamsianum	**Rossioglossum williamsianum**
Oncidium anceps	**Miltonia flava**
Oncidium cariniferum	**Miltonioides carinifera**
Oncidium clowesii	**Miltonia clowesii**
Oncidium flavescens	**Miltonia flavescens**
Oncidium insleayi	**Rossioglossum insleayi**
Oncidium karwinskii	**Miltonioides karwinskii**
Oncidium laeve	**Miltonioides laevis**
Onciduim leopardinum	**Rossioglossum splendens** var. **leopardinum**
Oncidium oviedomotae	**Miltonioides oviedomotae**
Oncidium reichenheimii	**Miltonioides reichenheimii**
Oncidium regnellii	**Miltonia regnellii**
Oncidium schroederianum	**Miltonioides schroederiana**
Oncidium spectabile	**Miltonia spectabilis**
Oncidium stellatum	**Miltonia flavescens**
Orchis lanigera	**Rhynchostylis retusa**
Orchis mauritiana	**Angraecum mauritianum**
Orchis triplicata	**Calanthe triplicata**
Paphiopedilum cothurnum	**Catasetum trulla**
Paphiopedilum socco	**Catasetum trulla**
Paracalanthe reflexa	**Calanthe reflexa**
Paracalanthe tricarinata	**Calanthe tricarinata**
Pectinaria thouarsii	**Angraecum pectinatum**
Phaius actinomorphus	**Calanthe actinomorpha**
Phaius calathoides	**Calanthe halconensis**
Phaius epiphyticus	**Calanthe densiflora**
Phaius vestitus	**Calanthe vestita**
Plectrelminthus spiculatus	**Aerangis spiculata**
Pogonia foliosa	**Bletilla foliosa**
Porphyrodesme elongata	**Renanthera elongata**

***For explanation see page 3, point 6**
***Voir les explications page 12, point 6**
***Para mayor explicación, véase la página 22, point 6**

ALL NAMES	ACCEPTED NAME
Preptanthe rubens	**Calanthe rubens**
Preptanthe vestita	**Calanthe vestita**
Preptanthe villosa	**Calanthe vestita**
Pteroceras caligare	**Vanda lilacina**
Radinocion flexuosa	**Aerangis flexuosa**
Renanthera alba	**Arachnis alba**
Renanthera amabilis	
Renanthera angustifolia	**Renanthera matutina**
Renanthera annamensis	
Renanthera arachnites	**Arachnis flosaeris**
Renanthera auyongi	**Renantherella histrionica**
Renanthera bella	
Renanthera bilinguis	**Arachnis labrosa**
Renanthera citrina	
Renanthera coccinea	
Renanthera coccinea var. *holttumii*	**Renanthera coccinea**
Renanthera edefeldtii	
Renanthera elongata	
Renanthera evrardii	**Arachnis annamensis**
Renanthera flosaeris	**Arachnis flosaeris**
Renanthera hennisiana	**Renanthera annamensis**
Renanthera histrionica	**Renantherella histrionica**
Renanthera hookeriana	**Arachnis hookeriana**
Renanthera imschootiana	
Renanthera isosepala	
Renanthera leptantha	**Arachnis labrosa**
Renanthera labrosa	**Arachnis labrosa**
Renanthera lowei	**Dimorphorchis lowii**
Renanthera maincathi	**Arachnis maingayi**
Renanthera maingayi	**Arachnis maingayi**
Renanthera matutina Lindl.	
Renanthera matutina auct. non (Blume) Lindl.	**Renanthera elongata**
Renanthera matutina Lindl. sensu Rolfe	**Renanthera imschootiana***
Renanthera matutina auct. non Lindl.	**Renanthera isosepala**
Renanthera micrantha	**Renanthera elongata**
Renanthera moluccana	
Renanthera monachica	
Renanthera moschifera	**Arachnis flosaeris**
Renanthera papilio	**Renanthera imschootiana**
Renanthera philippinensis	
Renanthera pulchella	**Renanthera annamensis**
Renanthera ramuana	**Sarcochilus moorei**
Renanthera rohaniana	**Arachnis hookeriana**
Renanthera sarcanthoides	**Porphyrodesme sarcanthoides**
Renanthera storiei	
Renanthera storiei forma *citrina*	**Renanthera storiei**
Renanthera storiei var. *philippensis*	**Renanthera philippensis**
Renanthera striata	
Renanthera sulingi	
Renanthera sulingii	**Arachnis sulingii**
Renanthera trichoglottis	**Vanda hastifera**
Renantherella histrionica	
Rhaphidorhynchus articulatus	**Aerangis articulata**
Rhaphidorhynchus batesii	**Aerangis arachnopus**
Rhaphidorhynchus bilobus	**Aerangis biloba**
Rhaphidorhynchus citratus	**Aerangis citrata**
Rhaphidorhynchus curnowianus	**Aerangis curnowiana**

***For explanation see page 3, point 6**
***Voir les explications page 12, point 6**
***Para mayor explicación, véase la página 22, point 6**

Part I: All Names / Tous les Noms / Todos los Nombres

ALL NAMES	ACCEPTED NAME
Rhaphidorhynchus fastuosus	**Aerangis fastuosa**
Rhaphidorhynchus kotschyi	**Aerangis kotschyana**
Rhaphidorhynchus luteo-albus	**Aerangis luteo-alba**
Rhaphidorhynchus modestus	**Aerangis modesta**
Rhaphidorhynchus modestus var. *sanderianus*	**Aerangis modesta**
Rhaphidorhynchus rohlfsianus	**Aerangis brachycarpa**
Rhaphidorhynchus spiculatus	**Aerangis spiculata**
Rhaphidorhynchus stylosus Finet	**Aerangis cryptodon**
Rhaphidorhynchus stylosus (Rolfe) Finet	**Aerangis stylosa***
Rhaphidorhynchus umbonatus	**Aerangis fuscata**
Rhynchostylis coelestis	
Rhynchostylis densiflora	**Rhynchostylis gigantea**
Rhynchostylis gigantea	
Rhynchostylis gigantea ssp. **violacea**	
Rhynchostylis gigantea subvar. *petotiana*	**Rhynchostylis gigantea**
Rhynchostylis gigantea var. *harrisoniana*	**Rhynchostylis gigantea**
Rhynchostylis gurwalica	**Rhynchostylis retusa**
Rhynchostylis guttata	**Rhynchostylis retusa**
Rhynchostylis hirsutus	**Chaerophyllum hirsutum**
Rhynchostylis latifolia	**Schoenorchis latifolia**
Rhynchostylis papillosa	**Acampe papillosa**
Rhynchostylis praemorsa	**Rhynchostylis retusa**
Rhynchostylis retusa	
Rhynchostylis retusa ssp. *macrostachya*	**Rhynchostylis retusa**
Rhynchostylis violacea Rchb.f.	**Rhynchostylis gigantea** ssp. **violacea**
Rhynchostylis violacea auct. non Rchb.f.	**Rhynchostylis retusa**
Rhynchostylis violacea ssp. *berkeleyi*	**Rhynchostylis retusa**
Rossioglossum grande	
Rossioglossum grande var. **aureum**	
Rossioglossum insleayi	
Rossioglossum powellii	
Rossioglossum schlieperianum	
Rossioglossum schlieperianum var. **flavidum**	
Rossioglossum splendens	
Rossioglossum splendens var. **imschootianum**	
Rossioglossum splendens var. **leopardinum**	
Rossioglossum splendens var. **pantherinum**	
Rossioglossum williamsianum	
Saccolabium albolineatum	**Rhynchostylis gigantea**
Saccolabium ampullaceum	**Ascocentrum ampullaceum**
Saccolabium aurantiacum	**Ascocentrum aurantiacum**
Saccolabium berkeleyi	**Rhynchostylis retusa**
Saccolabium blumei	**Rhynchostylis retusa**
Saccolabium blumei ssp. *major*	**Rhynchostylis retusa**
Saccolabium blumei var. *russelianum*	**Rhynchostylis retusa**
Saccolabium coeleste	**Rhynchostylis coelestis**
Saccolabium coriaceum	**Angraecum coriaceum**
Saccolabium curvifolium Lindl.	**Ascocentrum curvifolium**
Saccolabium curvifolium auct. non Lindl.	**Ascocentrum miniatum**
Saccolabium furcatum	**Rhynchostylis retusa**
Saccolabium giganteum	**Rhynchostylis gigantea**
Saccolabium gurwalicum	**Rhynchostylis retusa**
Saccolabium guttatum	**Rhynchostylis retusa**
Saccolabium harrisonianum	**Rhynchostylis gigantea**
Saccolabium heathii	**Rhynchostylis retusa**
Saccolabium hendersonianum	**Ascocentrum hendersoniana**
Saccolabium himalaicum	**Ascocentrum himalaicum**

***For explanation see page 3, point 6**
***Voir les explications page 12, point 6**
***Para mayor explicación, véase la página 22, point 6**

ALL NAMES	ACCEPTED NAME
Saccolabium holfordianum	**Rhynchostylis retusa**
Saccolabium littorale	**Rhynchostylis retusa**
Saccolabium macrostachyum	**Rhynchostylis retusa**
Saccolabium microphyton	**Angraecum tenellum**
Saccolabium miniatum	**Ascocentrum miniatum**
Saccolabium praemorsum	**Rhynchostylis retusa**
Saccolabium pumilum	**Ascocentrum pumilum**
Saccolabium reflexum	**Renanthera elongata**
Saccolabium retusum	**Rhynchostylis retusa**
Saccolabium rheedii	**Rhynchostylis retusa**
Saccolabium rubrum	**Ascocentrum rubrum**
Saccolabium spicatum	**Rhynchostylis retusa**
Saccolabium squamatum	**Angraecum bracteosum**
Saccolabium striatum	**Angraecum striatum**
Saccolabium turneri	**Rhynchostylis retusa**
Saccolabium violaceum auct. non Rchb.f.	**Rhynchostylis gigantea**
Saccolabium violaceum Rchb.f.	**Rhynchostylis gigantea** ssp. **violacea**
Sarcanthopsis muelleri	**Vandopsis muelleri**
Sarcanthus guttatus	**Rhynchostylis retusa**
Sarcochilus caligaris	**Vanda lilacina**
Stauropsis batemanii	**Vandopsis lissochiloides**
Stauropsis chinensis	**Vandopsis gigantea**
Stauropsis gigantea (Lindl.) Benth. & Hook.f.	**Vandopsis gigantea***
Stauropsis gigantea auct. non (Lindl.) Benth. & Hook.f.	**Vandopsis lissochiloides**
Stauropsis lissochiloides	**Vandopsis lissochiloides**
Stauropsis polyantha	**Vandopsis undulata**
Stauropsis shanica	**Vandopsis shanica**
Stauropsis undulata	**Vandopsis undulata**
Styloglossum nervosum	**Calanthe pulchra**
Taprobanea spathulata	**Vanda spathulata**
Trudelia alpina	**Vanda alpina**
Trudelia chlorosantha	**Vanda chlorosantha**
Trudelia cristata	**Vanda cristata**
Trudelia griffithii	**Vanda griffithii**
Trudelia pumila	**Vanda pumila**
Tulexis bicolor	**Brassavola tuberculata**
Vanda alpina	
Vanda amesiana	**Holcoglossum amesianum**
Vanda amiensis	**Vanda lamellata**
Vanda arbuthnotiana	
Vanda arcuata	
Vanda batemanii	**Vandopsis lissochiloides**
Vanda bensonii	
Vanda bicaudata	**Diploprora championii**
Vanda bicolor	
Vanda bidupensis	
Vanda × boumaniae	
Vanda boxallii	**Vanda lamellata**
Vanda boxallii ssp. *cobbiana*	**Vanda lamellata**
Vanda brunnea Rchb.f.	
Vanda brunnea auct. non Rchb.f.	**Vanda liouvillei**
Vanda cathcarti	**Esmerelda cathcartii**
Vanda celebica	
Vanda × charlesworthtii	
Vanda chlorosantha	
Vanda clarkei	**Esmerelda clarkei**

***For explanation see page 3, point 6**
***Voir les explications page 12, point 6**
***Para mayor explicaciôn, véase la página 22, point 6**

ALL NAMES	ACCEPTED NAME
Vanda clitellaria	**Vanda lamellata**
Vanda coerulea	
Vanda coerulea var. *concolor*	**Vanda coerulea**
Vanda coerulea ssp. *hennisiane*	**Vanda coerulea**
Vanda coerulea ssp. *sanderae*	**Vanda coerulea**
Vanda coerulescens Griff.	
Vanda coerulescens Lindl.	**Vanda coerulea***
Vanda coerulescens ssp. **boxallii**	
Vanda concolor	
Vanda × confusa	
Vanda congesta	**Acampe congesta**
Vanda crassiloba	
Vanda cristata	
Vanda cruenta	
Vanda cumingii	**Vanda lamellata**
Vanda dearei	
Vanda denevei	**Paraphalaenopsis denevei**
Vanda denisoniana	
Vanda denisoniana var. *hebraica*	**Vanda brunnea**
Vanda denisoniana var. *tessellata*	**Vanda denisoniana**
Vanda densiflora	**Rhynchostylis gigantea**
Vanda densiflora var. *petotiana*	**Rhynchostylis gigantea**
Vanda devoogtei	
Vanda doritoides	**Ornithochilus delavayi**
Vanda esquirolei	**Vanda concolor**
Vanda falcata	**Neofinetia falcatum**
Vanda fasciata	**Acampe wightiana**
Vanda fimbriata	**Gastrochilus acaulis**
Vanda flabellata	**Aerides flabellata**
Vanda flavobrunnea	
Vanda foetida	
Vanda furva	**Vanda concolor**
Vanda fuscoviridis	**Vanda concolor**
Vanda gibbsiae	**Vanda hastifera**
Vanda gigantea	**Vandopsis gigantea**
Vanda griffithii	
Vanda guangxiensis	**Vanda concolor**
Vanda guibertii	**Staurochilus guibertii**
Vanda hainanensis	**Rhynchostylis gigantea**
Vanda hastifera	
Vanda hastifera var. *gibbsiae*	**Vanda hastifera**
Vanda helvola	
Vanda henryi	**Vanda brunnea**
Vanda hindsii Lindl.	
Vanda hindsii Benth. non Lindl.	**Vanda tricolor***
Vanda hookeri	**Papilionanthe hookeriana**
Vanda hookeriana	**Papilionanthe hookeriana**
Vanda insignis	
Vanda insignis var. *schroederiana*	**Vanda insignis**
Vanda jainii	
Vanda javierae	
Vanda kimballiana	**Holcoglossum kimballianum**
Vanda kwangtungensis	**Vanda concolor**
Vanda lamellata	
Vanda lamellata forma *flava*	**Vanda lamellata**
Vanda lamellata var. *boxallii*	**Vanda lamellata**
Vanda lamellata var. *calayana*	**Vanda lamellata**

***For explanation see page 3, point 6**
***Voir les explications page 12, point 6**
***Para mayor explicación, véase la página 22, point 6**

ALL NAMES	ACCEPTED NAME
Vanda lamellata var. *cobbiana*	**Vanda lamellata**
Vanda lamellata var. *remediosae*	**Vanda lamellata**
Vanda laotica	**Vanda lilacina**
Vanda leucostele	
Vanda lilacina	
Vanda limbata	
Vanda lindeni	
Vanda lindenii	**Vanda concolor**
Vanda lindleyana	**Vandopsis gigantea**
Vanda liouvillei	
Vanda lissochiloides	**Vandopsis lissochiloides**
Vanda lombokensis	
Vanda longifolia	**Acampe rigida**
Vanda lowei	**Arachnis lowei**
Vanda lowii	**Dimorphorchis lowii**
Vanda luzonica	
Vanda masperoae	**Papilionanthe pedunculata**
Vanda merrillii	
Vanda merrillii var. *immaculata*	**Vanda merrillii**
Vanda merrillii var. *rotorii*	**Vanda merrillii**
Vanda micholitzii	**Vanda denisoniana**
Vanda moorei	
Vanda muelleri	**Vandopsis muelleri**
Vanda multiflora	**Acampe rigida**
Vanda nasughuana	**Vanda lamellata**
Vanda obliqua	**Gastrochilus obliquum**
Vanda paniculata	**Cleisostoma paniculatum**
Vanda parishii	**Hygrochilus parishii**
Vanda parishii ssp. *marriottiana*	**Hygrochilus parishii**
Vanda parishii ssp. *purpurea*	**Hygrochilus parishii**
Vanda parviflora auct. non Lindl.	**Vanda lilacina**
Vanda parviflora Lindl.	**Vanda testacea**
Vanda parviflora var. *albiflora*	**Vanda lilacina**
Vanda pauciflora	
Vanda peduncularis	**Cottonia peduncularis**
Vanda petersiana	**Vanda stangeana**
Vanda pseudo-caerulescens	**Rhynchostylis coelestis**
Vanda pulchella	**Gastrochilus acaulis**
Vanda pumila	
Vanda punctata	
Vanda pusilla	**Trichoglottis pusilla**
Vanda recurva	**Cleisostoma rostratum**
Vanda roeblingiana	
Vanda rostrata	**Cleisostoma rostratum**
Vanda roxburghii	**Vanda tessellata**
Vanda roxburgii var. *unicolor*	**Vanda concolor**
Vanda rupestris	**Holcoglossum rupestre**
Vanda sanderiana	
Vanda sanderiana var. *albata*	**Vanda sanderiana**
Vanda sanderiana var. *froebelliana*	**Vanda sanderiana**
Vanda sanderiana var. *labello-viridi*	**Vanda sanderiana**
Vanda saprophytica	**Holcoglossum saprophyticum**
Vanda saxatilis	**Vanda lindeni**
Vanda scandens	
Vanda scripta	**Grammatophyllum speciosum**
Vanda simondii	**Cleisostoma simondii**
Vanda spathulata (L.) Spreng.	

***For explanation see page 3, point 6**
***Voir les explications page 12, point 6**
***Para mayor explicación, véase la página 22, point 6**

ALL NAMES	ACCEPTED NAME
Vanda spathulata auct. non (L.) Spreng.	**Vanda testacea***
Vanda stangeana	
Vanda stella	**Vanda concolor**
Vanda storiei	**Renanthera storiei**
Vanda striata	**Vanda cristata**
Vanda suaveolens	**Vanda tricolor**
Vanda suavis auct. non Lindl.	**Vanda denisoniana***
Vanda suavis F.Muell.	**Vanda hindsii**
Vanda suavis Lindl.	**Vanda tricolor** ssp. **suavis**
Vanda subconcolor	
Vanda subconcolor var. *disticha*	**Vanda subconcolor**
Vanda subulifolia	**Holcoglossum subulifolium**
Vanda sulingi	**Renanthera sulingi**
Vanda sulingii	**Arachnis sulingii**
Vanda sumatrana	
Vanda superba	**Vanda lamellata**
Vanda taiwaniana	
Vanda teres	**Papilionanthe teres**
Vanda teres var. *candida*	**Papilionanthe teres**
Vanda teretifolia	**Cleisostoma simondii**
Vanda tessellata	
Vanda tessellata var. *lutescens*	**Vanda tessellata**
Vanda tessellata var. *rufescens*	**Vanda tessellata**
Vanda tesselloides	**Vanda tessellata**
Vanda testacea	
Vanda thwaitesii	
Vanda trichorhiza	**Luisia trichorhiza**
Vanda tricolor Lindl.	
Vanda tricolor Ames non Lindl.	**Vanda luzonica**
Vanda tricolor auct. non Lindl.	**Vanda denisoniana***
Vanda tricolor forma *patersonii*	**Vanda tricolor**
Vanda tricolor ssp. **suavis**	
Vanda tricolor var. *patersonii*	**Vanda tricolor**
Vanda tricuspidata	**Papilionanthe tricuspidata**
Vanda truncata	**Vanda hindsii**
Vanda undulata	**Vandopsis undulata**
Vanda unicolor sensu IK	**Vanda lamellata**
Vanda unicolor sensu Christenson	**Vanda tessellata**
Vanda vandarum	**Papilionanthe vandarum**
Vanda vidalii	**Vanda lamellata**
Vanda viminea	**Acampe rigida**
Vanda violacea	**Rhynchostylis gigantea** ssp. **violacea**
Vanda vipanii	
Vanda vitellina	**Vanda testacea**
Vanda watsoni	**Holcoglossum subulifolium**
Vanda whiteana	**Vanda hindsii**
Vanda wightiana	**Acampe praemorsa**
Vanda wightii	
Vanda yamiensis	**Vanda lamellata**
Vandopsis beccarii	**Vandopsis muelleri**
Vandopsis breviscapa	**Arachnis breviscapa**
Vandopsis celebica	**Arachnis celebica**
Vandopsis chalmersiana	**Sarcanthopsis nagarensis**
Vandopsis chinensis	**Vandopsis gigantea**
Vandopsis curvata	**Sarcanthopsis curvata**
Vandopsis davisii	**Trichoglottis loheriana**
Vandopsis gigantea	

***For explanation see page 3, point 6**
***Voir les explications page 12, point 6**
***Para mayor explicación, véase la página 22, point 6**

ALL NAMES	ACCEPTED NAME
Vandopsis hansemannii	**Sarcanthopsis nagarensis**
Vandopsis imthurnii	**Arachnis beccarii var imthurnii**
Vandopsis kupperiana	**Trichoglottis luzonensis**
Vandopsis leytensis	**Trichoglottis fasciata**
Vandopsis lissochiloides	
Vandopsis longicaulis	**Arachnis longicaulis**
Vandopsis lowii	**Dimorphorchis lowii**
Vandopsis luchuensis	**Trichoglottis lushuensis**
Vandopsis muelleri	
Vandopsis nagarensis	**Sarcanthopsis nagarensis**
Vandopsis pantherina	**Sarcanthopsis pantherina**
Vandopsis parishii	**Hygrochilus parishii**
Vandopsis praealta	**Sarcanthus praealtus**
Vandopsis quaifei	**Sarcanthopsis nagarensis**
Vandopsis raymundi	**Sarcanthopsis nagarensis**
Vandopsis shanica	
Vandopsis undulata	
Vandopsis warocqueana	**Sarcanthopsis nagarensis**
Vandopsis woodfordii	**Sarcanthopsis nagarensis**
Zoduba masuca	**Calanthe sylvatica**

***For explanation see page 3, point 6**
***Voir les explications page 12, point 6**
***Para mayor explicación, véase la página 22, point 6**

PART II: ACCEPTED NAMES IN CURRENT USE
Ordered alphabetically on Accepted Names for the genera:

Aerangis, Angraecum, Ascocentrum, Bletilla, Brassavola, Calanthe, Catasetum, Miltonia, Miltonioides, Miltoniopsis, Renanthera, Renantherella, Rhynchostylis, Rossioglossum, Vanda and *Vandopsis*

DEUXIEME PARTIE: NOMS ACCEPTES D'USAGE COURANT
Par ordre alphabétique des noms acceptés pour les genre:

Aerangis, Angraecum, Ascocentrum, Bletilla, Brassavola, Calanthe, Catasetum, Miltonia, Miltonioides, Miltoniopsis, Renanthera, Renantherella, Rhynchostylis, Rossioglossum, Vanda et *Vandopsis*

PARTE II: NOMBRES ACEPTADOS UTILIZADOS NORMALMENTE
Presentados por orden alfabético: nombres aceptados para el genero:

Aerangis, Angraecum, Ascocentrum, Bletilla, Brassavola, Calanthe, Catasetum, Miltonia, Miltonioides, Miltoniopsis, Renanthera, Renantherella, Rhynchostylis, Rossioglossum, Vanda y *Vandopsis*

AERANGIS BINOMIALS IN CURRENT USE

AERANGIS BINOMES ACTUELLEMENT EN USAGE

AERANGIS BINOMIALES UTILIZADOS NORMALMENTE

Aerangis alcicornis (Rchb.f.) Garay
Aerangis calodictyon Summerh.
Aerangis lutambae Mansf.
Angorchis alcicornis (Rchb.f.) Kuntze
Angraecum alcicorne Rchb.f.

Distribution: Malawi, Mozambique, Tanzania (United Republic of)

Aerangis appendiculata (De Wild.) Schltr.
Aerangis pachyura sensu Morris non Schltr.
Mystacidium appendiculatum De Wild.

Distribution: Malawi, Mozambique, Zambia, Zimbabwe

Aerangis arachnopus (Rchb.f.) Schltr.
Aerangis batesii (Rolfe) Schltr.
Aerangis biloboides (De Wild.) Schltr.
Angorchis arachnopus (Rchb.f.) Kuntze
Angraecum arachnopus Rchb.f.
Angraecum batesii Rolfe
Angraecum biloboides De Wild.
Rhaphidorhynchus batesii (Rolfe) Finet

Distribution: Cameroon, Congo (The Democratic Republic of the), Gabon, Ghana

Aerangis articulata (Rchb.f.) Schltr.
Aerangis calligera (Rchb.f.) Garay
Aerangis stylosa sensu Perrier
Aerangis venusta Schltr.
Angorchis articulata (Rchb.f.) Kuntze
Angraecum articulatum Rchb.f.
Angraecum calligerum Rchb.f.
Angraecum descendens Rchb.f.
Angraecum venustum Schltr.
Rhaphidorhynchus articulatus (Rchb.f.) Poiss.

Distribution: Comoros (The), Madagascar

Aerangis biloba (Lindl.) Schltr.
Aerangis campyloplectron (Rchb.f.) Garay
Angorchis biloba (Lindl.) Kuntze
Angorchis campyloplectron (Rchb.f.) Kuntze
Angraecum apiculatum Hook.f.
Angraecum bilobum Lindl.
Angraecum campyloplectron Rchb.f.

Listrostachys biloba (Lindl.) Kraenzl.
Rhaphidorhynchus bilobus (Lindl.) Finet

Distribution: Cameroon, Côte d'Ivoire, Ghana, Guinea, Liberia, Nigeria, Senegal, Sierra Leone, Togo

Aerangis bouarensis Chiron

Distribution: Central African Republic (The)

Aerangis brachycarpa (A.Rich.) Dur. & Schinz
Aerangis carusiana (Severino) Garay
Aerangis flabellifolia Rchb.f.
Aerangis friesiorum sensu Tweedie non Schltr.
Aerangis rohlfsiana (Kraenzl.) Schltr.
Angorchis flabellifolia (Rchb.f.) Kuntze
Angraecum brachycarpum (A.Rich.) Rchb.f.
Angraecum bilobum sensu Engl. non Lindl.
Angraecum bilobum sensu Schweinf. non Lindl.
Angraecum carusianum Severino
Angraecum flabellifolium (Rchb.f.) Rolfe
Angraecum rohlfsianum Kraenzl.
Dendrobium brachycarpum A.Rich.
Rhaphidorhynchus rohlfsianus (Kraenzl.) Finet

Distribution: Angola, Ethiopia, Kenya, Tanzania (United Republic of), Uganda, Zambia

Aerangis calantha (Schltr.) Schltr.
Aerangis parvula Schltr.
Aerangis roseocalcarata (De Wild.) Schltr.
Aerangis sankuruensis (De Wild.) Schltr.
Angraecum calanthum Schltr.
Angraecum roseocalcaratum De Wild.
Angraecum sankuruense De Wild.

Distribution: Angola, Cameroon, Central African Republic (The), Congo (The Democratic Republic of the), Equatorial Guinea, Ghana, Tanzania (United Republic of), Uganda

Aerangis carnea J.Stewart

Distribution: Malawi, Tanzania (United Republic of)

Aerangis citrata (Thouars) Schltr.
Aerobion citratum (Thouars) Spreng.
Angorchis citrata (Thouars) Kuntze
Angraecum citratum Thouars
Rhaphidorhynchus citratus (Thouars) Finet

Distribution: Madagascar

Aerangis collum-cygni Summerh.
Aerangis compta Summerh.

Distribution: Cameroon, Central African Republic (The), Congo (The Democratic Republic of the), Tanzania (United Republic of), Uganda, Zambia

Aerangis concavipetala H.Perrier

Distribution: Madagascar

Aerangis confusa J.Stewart

Distribution: Kenya, Tanzania (United Republic of)

Aerangis coriacea Summerh.

Distribution: Kenya, Tanzania (United Republic of)

Aerangis cryptodon (Rchb.f.) Schltr.
Aerangis malmquistiana Schltr.
Angorchis cryptodon (Rchb.f.) Kuntze
Angraecum cryptodon Rchb.f.
Rhaphidorhynchus stylosus Finet

Distribution: Madagascar

Aerangis curnowiana (Finet) H.Perrier
Rhaphidorhynchus curnowianus Finet

Distribution: Madagascar

Aerangis decaryana H.Perrier

Distribution: Madagascar

Aerangis distincta J.Stewart & la Croix

Distribution: Malawi

Aerangis ellisii (B.S.Williams) Schltr.
Aerangis alata H.Perrier
Aerangis buyssonii God.-Leb.
Aerangis caulescens Schltr.
Aerangis cryptodon sensu H.Perrier
Aerangis platyphylla Schltr.
Angorchis ellisii (B.S.Williams) Kuntze
Angraecum buyssonii God.-Leb.
Angraecum dubuyssonii God.-Leb.

Part II: Aerangis

Angraecum ellisii B.S.Williams

Distribution: Madagascar

Aerangis ellisii var. **grandiflora** J.Stewart

Distribution: Madagascar

Aerangis fastuosa (Rchb.f.) Schltr.
Aerangis fastuosa ssp. *francoisii* H.Perrier
Aerangis fastuosa ssp. *grandidieri* H.Perrier
Aerangis fastuosa ssp. *maculata* H.Perrier
Aerangis fastuosa ssp. *rotundifolia* H.Perrier
Aerangis fastuosa ssp. *vondrozensis* H.Perrier
Angorchis fastuosa (Rchb.f.) Kuntze
Angraecum fastuosum Rchb.f.
Rhaphidorhynchus fastuosus (Rchb.f.) Finet

Distribution: Madagascar

Aerangis flexuosa (Ridl.) Schltr.
Aerangis elegans (Rolfe) Dandy
Aerangis henriquesiana (Rolfe) Schltr.
Angraecum elegans Rolfe
Angraecum flexuosum (Ridl.) Rolfe
Angraecum henriquesianum Rolfe
Radinocion flexuosa Ridl.

Distribution: Sao Tome and Principe

Aerangis fuscata (Rchb.f.) Schltr.
Aerangis monantha Schltr.
Aerangis umbonata (Finet) Schltr.
Angraecum fuscatum Rchb.f.
Rhaphidorhynchus umbonatus Finet

Distribution: Madagascar

Aerangis gracillima (Kraenzl.) J.Č.Arends & J.Stewart
Angraecum gracillimum Kraenzl.
Barombia gracillima (Kraenzl.) Schltr.

Distribution: Cameroon, Gabon

Aerangis gravenreuthii (Kraenzl.) Schltr.
Aeranthus gravenreuthii Kraenzl.
Angraecum ellisii var. *occidentale* Kraenzl.
Angraecum gravenreutii Kraenzl.
Angraecum stella Schltr.

Mystacidium gravenreuthii (Kraenzl.) Rolfe

Distribution: Cameroon, Equatorial Guinea, Tanzania (United Republic of)

Aerangis hologlottis (Schltr.) Schltr.
Angraecum hologlottis Schltr.

Distribution: Kenya, Mozambique, Sri Lanka, Tanzania (United Republic of)

Aerangis hyaloides (Rchb.f.) Schltr.
Aerangis pumilio Schltr.
Angorchis hyaloides (Rchb.f.) Kuntze
Angraecum hyaloides Rchb.f.

Distribution: Madagascar

Aerangis jacksonii J.Stewart

Distribution: Uganda

Aerangis kirkii (Rchb.f.) Schltr.
Aerangis biloba ssp. *kirkii* (Rchb.f.) Hawkes
Angraecum apiculatum ssp. *kirkii* (Rchb.f.) Rchb.f.
Angraecum bilobum ssp. *kirkii* Rchb.f.
Angraecum kirkii (Rchb.f.) Rolfe

Distribution: Kenya, Malawi, Mozambique, Tanzania (United Republic of)

Aerangis kotschyana (Rchb.f.) Schltr.
Aerangis grantii (Baker) Schltr.
Aerangis kotschyi (Rchb.f.) Rchb.f.
Angraecum grantii Baker
Angraecum kotschyanum Rchb.f.
Angraecum kotschyi Rchb.f.
Angraecum semipedale Rendle
Rhaphidorhynchus kotschyi (Rchb.f.) Finet

Distribution: Burundi, Central African Republic (The), Congo (The Democratic Republic of the), Ethiopia, Guinea, Kenya, Malawi, Mozambique, Nigeria, Rwanda, Sudan (The), Tanzania (United Republic of), Uganda, Zambia, Zimbabwe

Aerangis luteo-alba (Kraenzl.) Schltr.
Angraecum luteo-album Kraenzl.
Rhaphidorhynchus luteo-albus (Kraenzl.) Finet

Distribution: Cameroon, Central African Republic (The), Congo (The Democratic Republic of the), Ethiopia, Kenya, Tanzania (United Republic of), Uganda

Aerangis luteo-alba var. **luteo-alba**

Distribution: Congo (The Democratic Republic of the)

Aerangis luteo-alba var. **rhodosticta** (Kraenzl.) J.Stewart
Aerangis albido-rubra (De Wild.) Schltr.
Aerangis rhodosticta (Kraenzl.) Schltr.
Angorchis rhodosticta (Kraenzl.) Kuntze
Angraecum albido-rubrum De Wild.
Angraecum mirabile Hort non Schltr.
Angraecum rhodostictum Kraenzl.

Distribution: Cameroon, Central African Republic (The), Congo (The Democratic Republic of the), Ethiopia, Kenya, Tanzania (United Republic of), Uganda

Aerangis macrocentra (Schltr.) Schltr.
Aerangis clavigera H.Perrier
Angraecum macrocentrum Schltr.

Distribution: Madagascar

Aerangis maireae la Croix & J.Stewart

Distribution: Tanzania (United Republic of)

Aerangis megaphylla Summerh.
Aerangis phalaenopsis Schltr.

Distribution: Cameroon, Central African Republic (The), Equatorial Guinea

Aerangis modesta (Hook.f.) Schltr.
Aerangis crassipes Schltr.
Aerangis fastuosa ssp. *angustifolia* H.Perrier
Angorchis modesta (Hook.f.) Kuntze
Angraecum modestum Hook.f.
Angraecum sanderianum Rchb.f.
Rhaphidorhynchus modestus (Hook.f.) Finet
Rhaphidorhynchus modestus ssp. *sanderianus* Poiss.

Distribution: Comoros (The), Madagascar

Aerangis montana J.Stewart

Distribution: Malawi, Tanzania (United Republic of), Zambia

Aerangis mooreana (Rolfe ex Sander) P.J.Cribb & J.Stewart
Aerangis anjoanensis H.Perrier
Aerangis ikopana Schltr.

Angraecum mooreanum Rolfe ex Sander

Distribution: Comoros (The), Madagascar

Aerangis mystacidii (Rchb.f.) Schltr.
Aerangis mystacidioides Schltr.
Aerangis pachyura (Rolfe) Schltr.
Angraecum mystacidii Rchb.f.
Angraecum pachyurum Rolfe
Angraecum saundersiae Bolus

Distribution: Malawi, Mozambique, South Africa, Swaziland, Tanzania (United Republic of), Zambia, Zimbabwe

Aerangis oligantha Schltr.

Distribution: Malawi, Tanzania (United Republic of)

Aerangis pallida (W.Watson) Garay
Angraecum pallidum W.Watson

Distribution: Madagascar

Aerangis pallidiflora H.Perrier
Aerangis seegeri Senghas

Distribution: Madagascar

Aerangis × primulina (Rolfe) H.Perrier
Aerangis primulina (Rolfe) H.Perrier
Angraecum primulinum Rolfe

Distribution: Madagascar

Aerangis pulchella (Schltr.) Schltr.
Angraecum pulchellum Schltr.

Distribution: Madagascar

Aerangis punctata J.Stewart

Distribution: Madagascar

Aerangis rostellaris (Rchb.f.) H.Perrier
Aerangis avicularia (Rchb.f.) Schltr.
Aerangis buchlohii Senghas
Angraecum avicularium Rchb.f.

Angraecum rostellare Rchb.f.

Distribution: Comoros (The), Madagascar

Aerangis somalensis (Schltr.) Schltr.
Angraecum somalense Schltr.

Distribution: Ethiopia, Kenya, Malawi, South Africa, Tanzania (United Republic of), Zimbabwe

Aerangis spiculata (Finet) Sengas
Angraecum fuscatum sensu Carriere
Leptocentrum spiculatum (Finet) Schltr.
Plectrelminthus spiculatus (Finet) Summerh.
Rhaphidorhynchus spiculatus Finet

Distribution: Comoros (The), Madagascar

Aerangis splendida J.Stewart & la Croix
Aerangis kotschyana sensu Morris non Schltr.

Distribution: Malawi, Zambia

Aerangis stelligera Summerh.

Distribution: Cameroon, Central African Republic (The), Congo (The Democratic Republic of the)

Aerangis stylosa (Rolfe) Schltr.
Angraecum fournierae Andre
Aerangis fuscata sensu H.Perrier
Angraecum stylosum Rolfe
Rhaphidorhynchus stylosus (Rolfe) Finet

Distribution: Comoros (The), Madagascar

Aerangis thomsonii (Rolfe) Schltr.
Aerangis friesiorum Schltr.
Angraecum thomsonii Rolfe

Distribution: Kenya, Tanzania (United Republic of), Uganda

Aerangis ugandensis Summerh.

Distribution: Burundi, Congo (The Democratic Republic of the), Kenya, Uganda

Aerangis verdickii (De Wild.) Schltr.
Aerangis erythrurum (Kraenzl.) Garay
Aerangis kotschyana sensu Schelpe non Schltr.
Aerangis schliebenii nomen Mansf.

Angraecum augustum Rolfe
Angraecum erythrurum Kraenzl.
Angraecum verdickii De Wild.

Distribution: Angola, Congo (The Democratic Republic of the), Malawi, Mozambique, South Africa, Tanzania (United Republic of), Zambia, Zimbabwe

Aerangis verdickii ssp. **rusituensis** (Fibeck & Dare) la Croix & P.J.Cribb
Aerangis rusituensis Fibeck & Dare

Distribution: Zimbabwe

ANGRAECUM BINOMIALS IN CURRENT USE

ANGRAECUM BINOMES ACTUELLEMENT EN USAGE

ANGRAECUM BINOMIALES UTILIZADOS NORMALMENTE

Angraecum acutipetalum Schltr.

Distribution: Madagascar

Angraecum acutipetalum ssp. **analabeensis** H.Perrier

Distribution: Madagascar

Angraecum acutipetalum ssp. **ankeranae** H.Perrier

Distribution: Madagascar

Angraecum affine Schltr.
Angraecum ligulatum Summerh.

Distribution: Cameroon, Congo (The), Congo (The Democratic Republic of the), Equatorial Guinea, Uganda

Angraecum alleizettei Schltr.

Distribution: Madagascar

Angraecum aloifolium Hermans & P.J.Cribb

Distribution: Madagascar

Angraecum ambrense H.Perrier

Distribution: Madagascar

Angraecum amplexicaule Toill.-Gen. & Bosser

Distribution: Madagascar

Angraecum ampullaceum Bosser
Jumellea humberti H.Perrier

Distribution: Madagascar

Part II: Angraecum

Angraecum andasibeense H.Perrier

Distribution: Madagascar

Angraecum andringitranum Schltr.

Distribution: Madagascar

Angraecum angustipetalum Rendle
Angraecum boonei De Wild.

Distribution: Cameroon, Congo (The Democratic Republic of the), Gabon, Ghana, Malawi, Nigeria

Angraecum angustum (Rolfe) Summerh.
Mystacidium angustum Rolfe

Distribution: Nigeria

Angraecum ankeranense H.Perrier

Distribution: Madagascar

Angraecum aporoides Summerh.
Angraecum distichum ssp. *grandifolium* (De Wild.) Summerh.
Mystacidium distichum ssp. *grandifolium* De Wild.

Distribution: Cameroon, Congo (The Democratic Republic of the), Equatorial Guinea, Nigeria, Sao Tome and Principe

Angraecum appendiculoides Schltr.

Distribution: Madagascar

Angraecum astroarche Ridl.
Mystacidium astroarche (Ridl.) Rolfe

Distribution: Sao Tome and Principe

Angraecum aviceps Schltr.

Distribution: Madagascar

Angraecum bancoense Van der Burg

Distribution: Cameroon, Congo (The), Côte d'Ivoire

Angraecum baronii (Finet) Schltr.
Angraecum dichaeoides Schltr.
Macroplectrum baronii Finet

Distribution: Madagascar

Angraecum bemarivoense Schltr.

Distribution: Madagascar

Angraecum bicallosum H.Perrier

Distribution: Madagascar

Angraecum birrimense Rolfe

Distribution: Cameroon, Côte d'Ivoire, Ghana, Liberia, Nigeria, Sierra Leone

Angraecum borbonicum Bosser

Distribution: Mauritius, Reunion

Angraecum brachyrhopalon Schltr.

Distribution: Madagascar

Angraecum bracteosum Balf.f.& S.Moore
Listrostachys bracteosa (Balf.f. & S.Moore) Rolfe
Saccolabium squamatum Frapp. ex Cordem.

Distribution: Reunion

Angraecum breve Schltr.

Distribution: Madagascar

Angraecum brevicornu Summerh.

Distribution: Tanzania (United Republic of)

Angraecum cadetii Bosser

Distribution: Mauritius, Reunion

Angraecum calceolus Thouars
Aeranthus calceolus (Thouars) Baker

Part II: Angraecum

Aerobion calceolus Spreng.
Angraecum anocentrum Schltr.
Angraecum carpophorum Thouars
Angraecum paniculatum Frapp. ex Cordem.
Angraecum patens Frapp.
Angraecum rhopaloceras Schltr.
Epidorchis calceolus (Thouars) Kuntze
Macroplectrum calceolus (Thouars) Finet
Mystacidium calceolus (Thouars) Cordem.

Distribution: Comoros (The), Madagascar, Mauritius, Mozambique, Reunion, Seychelles

Angraecum caricifolium H.Perrier
Angraecum carifolium H.Perrier (sphalm)

Distribution: Madagascar

Angraecum caulescens Thouars
Aerobion caulescens (Thouars) Spreng.
Epidorchis caulescens (Thouars) Kuntze
Mystacidium caulescens (Thouars) Ridl.

Distribution: Madagascar, Mauritius, Reunion

Angraecum chaetopodum Schltr.

Distribution: Madagascar

Angraecum chamaeanthus Schltr.

Distribution: Kenya, Malawi, Mozambique, South Africa, Tanzania (United Republic of), Zimbabwe

Angraecum chermezoni H.Perrier

Distribution: Madagascar

Angraecum chimanimaniense G.Will.

Distribution: Zimbabwe

Angraecum chloranthum Schltr.

Distribution: Madagascar

Angraecum cilaosianum (Cordem.) Schltr.
Mystacidium cilaosianum Cordem.

Distribution: Reunion

Angraecum claessensii De Wild.

Distribution: Congo (The Democratic Republic of the), Liberia, Nigeria

Angraecum clavigerum Ridl.
Angorchis clavigera (Ridl.) Kuntze
Monixus clavigera (Ridl.) Finet

Distribution: Madagascar

Angraecum compactum Schltr.

Distribution: Madagascar

Angraecum compressicaule H.Perrier

Distribution: Madagascar

Angraecum conchiferum Lindl.
Angorchis conchifera (Lindl.) Kuntze
Angraecum scabripes Kraenzl.
Angraecum verrucosum Rendle
Mystacidium verrucosum (Rendle) Rolfe

Distribution: Kenya, Malawi, Mozambique, South Africa, Tanzania (United Republic of), Zimbabwe

Angraecum cordemoyi Schltr.
Mystacidium striatum Cordem.

Distribution: Reunion

Angraecum coriaceum (Thunb. ex Sw.) Schltr.
Aerides coriaceum Thunb. ex Sw.
Epidendrum coriaceum (Thunb. ex Sw.) Poir.
Gastrochilus coriaceus (Thunb. ex Sw.) Kuntze
Limodorum coriaceum Thunb. ex Sw.
Saccolabium coriaceum (Thunb. ex Sw.) Lindl.

Distribution: Madagascar

Part II: Angraecum

Angraecum cornigerum Cordem.

Distribution: Reunion

Angraecum cornucopiae H.Perrier

Distribution: Madagascar

Angraecum corynoceras Schltr.

Distribution: Madagascar

Angraecum costatum Frapp. ex Cordem.
Angraecum baronii (Finet) Schltr.
Macroplectrum costatum Finet
Mystacidium costatum Cordem.

Distribution: Mauritius, Reunion

Angraecum coutrixii Bosser

Distribution: Madagascar

Angraecum crassifolium (Cordem.) Schltr.
Mystacidium crassifolium Cordem.

Distribution: Reunion

Angraecum crassum Thouars
Aerobion crassum (Thouars) Spreng.
Angorchis crassa (Thouars) Kuntze
Angraecum crassiflorum H.Perrier
Angraecum sarcodanthum Schltr.

Distribution: Madagascar

Angraecum cribbianum Szl. & Olzs.

Distribution: Gabon

Angraecum cucullatum Thouars
Aerobion cucullatum Spreng.
Angorchis cucullata Kuntze
Angorchis fragrans Kuntze
Macroplectrum cucullatum Finet

Distribution: Reunion

Angraecum cultriforme Summerh.

Distribution: Kenya, Malawi, Mozambique, South Africa, Tanzania (United Republic of), Zambia, Zimbabwe

Angraecum curnowianum (Rchb.f.) Dur. & Schinz
Aeranthus curnowianus Rchb.f.
Angorchis curnowiana (Rchb.f.) Kuntze
Angraecum suarezense Toill.-Gen. & Bosser
Angraecum subcordatum (H.Perrier) Bosser
Jumellea curnowiana (Rchb.f.) Schltr.
Jumellea subcordata H.Perrier
Mystacidium curnowianus (Rchb.f.) Rolfe

Distribution: Madagascar

Angraecum curvicalcar Schltr.

Distribution: Madagascar

Angraecum curvicaule Schltr.

Distribution: Madagascar

Angraecum curvipes Schltr.

Distribution: Cameroon

Angraecum danguyanum H.Perrier

Distribution: Madagascar

Angraecum dasycarpum Schltr.

Distribution: Madagascar

Angraecum dauphinense (Rolfe) Schltr.
Mystacidium dauphinense Rolfe

Distribution: Madagascar

Angraecum decaryanum H.Perrier

Distribution: Madagascar

Part II: Angraecum

Angraecum decipiens Summerh.

Distribution: Kenya, Tanzania (United Republic of)

Angraecum dendrobiopsis Schltr.

Distribution: Madagascar

Angraecum didieri (Baill. ex Finet) Schltr.
Macroplectrum didieri Baill. ex Finet

Distribution: Madagascar

Angraecum distichum Lindl.
Angraecum imbricatum (Sw.) Schltr.
Mystacidium distichum (Lindl.) Pfitzer

Distribution: Angola, Benin, Cameroon, Central African Republic (The), Congo (The), Congo (The Democratic Republic of the), Côte d'Ivoire, Gabon, Ghana, Guinea, Liberia, Nigeria, Sierra Leone, Uganda

Angraecum divaricatum Frapp.

Distribution: Mauritius, Reunion

Angraecum dives Rolfe

Distribution: Kenya, Tanzania (United Republic of), Yemen

Angraecum dollii Senghas

Distribution: Madagascar

Angraecum doratophyllum Summerh.

Distribution: Sao Tome and Principe

Angraecum drouhardii H.Perrier

Distribution: Madagascar

Angraecum dryadum Schltr.

Distribution: Madagascar

Angraecum eburneum Bory
Angorchis eburnea (Bory) Kuntze
Angraecum eburneum ssp. *typicum* H.Perrier
Angraecum eburneum ssp. *virens* Hook.
Angraecum virens Lindl.
Limodorum eburneum (Bory) Willd.

Distribution: Comoros (The), Madagascar, Mauritius, Reunion, Seychelles

Angraecum eburneum ssp. **giryamae** (Rendle) Senghas & Cribb
Angraecum giryamae Rendle

Distribution: Kenya, Tanzania (United Republic of)

Angraecum eburneum ssp. **superbum** Thouars
Aerobion superbum Thouars
Angorchis brongniartiana (Rchb.f. ex Linden) Kuntze
Angorchis superba (Thouars) Kuntze
Angraecum brongniartianum Rchb.f. ex Linden
Angraecum comorense Kraenzl. non (Rchb.f.) Finet
Angraecum eburneum ssp. *brongniartianum* (Rchb.f. ex Linden) Schltr.
Angraecum superbum Thouars
Angraecum voeltzkowianum Kraenzl.

Distribution: Comoros (The), Madagascar, Seychelles

Angraecum eburneum var. **longicalcar** Bosser
Angraecum eburneum ssp. *longicalcar* Bosser
Angraecum longicalcar (Bosser) Senghas

Distribution: Madagascar

Angraecum eburneum ssp. **xerophilum** H.Perrier

Distribution: Madagascar

Angraecum egertonii Rendle

Distribution: Gabon, Nigeria

Angraecum eichlerianum Kraenzl.
Angraecum arnoldianum De Wild.

Distribution: Angola, Cameroon, Congo (The Democratic Republic of the), Gabon, Nigeria

Angraecum eichlerianum var. **curvicalcaratum** Szl. & Olzs.

Distribution: Cameroon, Gabon

Part II: Angraecum

Angraecum elephantinum Schltr.

Distribution: Madagascar

Angraecum elliotii Rolfe

Distribution: Madagascar

Angraecum equitans Schltr.

Distribution: Madagascar

Angraecum erectum Summerh.

Distribution: Kenya, Tanzania (United Republic of), Uganda, Zambia

Angraecum evrardianum Geerinck

Distribution: Burundi

Angraecum expansum Thouars

Distribution: Reunion

Angraecum expansum ssp. **inflatum** Cordem.

Distribution: Reunion

Angraecum falcifolium Bosser

Distribution: Madagascar

Angraecum ferkoanum Schltr.

Distribution: Madagascar

Angraecum filicornu Thouars
Aeranthes thouarsii S.Moore
Aerobion filicornu (Thouars) Spreng.

Distribution: Madagascar, Mauritius, Reunion

Angraecum firthii Summerh.

Distribution: Cameroon, Kenya, Uganda

Angraecum flavidum Bosser

Distribution: Madagascar

Angraecum floribundum Bosser

Distribution: Madagascar

Angraecum florulentum Rchb.f.

Distribution: Comoros (The), Madagascar

Angraecum gabonense Summerh.

Distribution: Congo (The Democratic Republic of the), Gabon

Angraecum geniculatum G.Williamson

Distribution: Zambia

Angraecum germinyanum Hook.f.
Angraecum arachnites Schltr.
Angraecum bathiei Schltr.
Angraecum bathiei ssp. *peracuminatum* H.Perrier
Angraecum conchoglossum Schltr.
Angraecum ramosum auct. non Thouars, H.Perrier
Angraecum ramosum ssp. *bidentatum* H.Perrier
Angraecum ramosum ssp. *typicum* H.Perrier
Angraecum ramosum ssp. *typicum* var. *arachnites* (Schltr.) H.Perrier
Angraecum ramosum ssp. *typicum* var. *bathiei* (Schltr.) H.Perrier
Angraecum ramosum ssp. *typicum* var. *conchoglossum* (Schltr.) H.Perrier
Angraecum ramosum ssp. *typicum* var. *peracuminatum* H.Perrier
Mystacidium germinyanum (Hook.f.) Rolfe

Distribution: Comoros (The), Madagascar, Reunion

Angraecum guillauminii H.Perrier

Distribution: Madagascar

Angraecum hermannii (Cordem.) Schltr.
Mystacidium hermanni Cordem.

Distribution: Reunion

Angraecum humbertii H.Perrier

Distribution: Madagascar

Part II: Angraecum

Angraecum humblotianum (Finet) Schltr.
Angraecum abietinum Schltr.
Angraecum finetianum Schltr.
Angraecum humblotii (Finet) Summerh.
Macroplectrum humblotii Finet

Distribution: Madagascar

Angraecum humile Summerh.

Distribution: Kenya, Tanzania (United Republic of), Zimbabwe

Angraecum huntleyoides Schltr.

Distribution: Madagascar

Angraecum imerinense Schltr.

Distribution: Madagascar

Angraecum implicatum Thouars
Aerobion implicatum (Thouars) Spreng.
Angorchis implicata (Thouars) Kuntze
Angraecum verruculosum Frapp. ex Cordem.
Macroplectrum implicatum (Thouars) Finet

Distribution: Madagascar, Reunion

Angraecum inapertum Thouars
Aerobion inapertum (Thouars) Spreng.
Epidorchis inaperta (Thouars) Kuntze
Mystacidium inapertum (Thouars) Ridl.

Distribution: Madagascar, Mauritius, Reunion

Angraecum infundibulare Lindl.
Angorchis infundibularis (Lindl.) Kuntze
Mystacidium infundibulare (Lindl.) Rolfe

Distribution: Cameroon, Congo (The Democratic Republic of the), Ethiopia, Kenya, Nigeria, Rwanda, Sao Tome and Principe, Uganda

Angraecum keniae Kraenzl.
Mystacidium keniae (Kraenzl.) Rolfe

Distribution: Kenya

Angraecum kraenzlinianum H.Perrier
Aeranthes englerianus Kraenzl.
Angraecum englerianum (Kraenzl.) Schltr. non Kranzl.
Angraecum robustum Kraenzl. non Schltr.

Distribution: Madagascar

Angraecum laggiarae Schltr.

Distribution: Madagascar

Angraecum lecomtei H.Perrier

Distribution: Madagascar

Angraecum leonis (Rchb.f.) Andre
Aeranthes leonis Rchb.f.
Aeranthus leonis Rchb.f.
Angraecum humblotii Rchb.f. ex Rolfe
Angraecum leonii Andre
Macroplectrum leonis (Rchb.f.) Finet
Mystacidium leonis (Rchb.f.) Rolfe

Distribution: Comoros (The), Madagascar

Angraecum letouzeyi Bosser

Distribution: Madagascar

Angraecum linearifolium Garay
Angraecum palmiforme H.Perrier

Distribution: Madagascar

Angraecum lisowskianum Szl. & Olzs.

Distribution: Equatorial Guinea

Angraecum litorale Schltr.

Distribution: Madagascar

Angraecum longicaule H.Perrier

Distribution: Madagascar

Angraecum longinode Frapp. ex Cordem.
Mystacidium longinode Cordem.

Distribution: Reunion

Angraecum macilentum Frapp.

Distribution: Reunion

Angraecum madagascariense (Finet) Schltr.
Macroplectrum madagascariense Finet

Distribution: Madagascar

Angraecum magdalenae Schltr.

Distribution: Madagascar

Angraecum magdalenae var. **latilabellum** Bosser

Distribution: Madagascar

Angraecum mahavavense H.Perrier

Distribution: Madagascar

Angraecum marii Geerinck
Eggelingia ligulifolia Summerh.

Distribution: Rwanda

Angraecum mauritianum (Poir.) Frapp.
Aeranthes gladiifolius (Thouars) Rchb.f.
Aeranthus gladiifolius (Thouars) Rchb.f.
Aerobion gladiifolium (Thouars) Spreng.
Angorchis gladiifolia (Thouars) Kuntze
Angraecum gladiifolium Thouars
Macroplectrum gladiifolium (Thouars) Pfitzer ex Finet
Mystacidium gladiifolium (Thouars) Rolfe
Mystacidium mauritianum (Lam.) T.Durand & Schinz
Orchis mauritiana Poir.

Distribution: Madagascar, Mauritius, Reunion

Angraecum meirax (Rchb.f.) H.Perrier
Aeranthus meirax Rchb.f.
Jumellea meirax (Rchb.f.) Schltr.

Macroplectrum meirax (Rchb.f.) Finet

Distribution: Comoros (The), Madagascar

Angraecum melanostictum Schltr.

Distribution: Madagascar

Angraecum metallicum Sander

Distribution: Madagascar

Angraecum microcharis Schltr.

Distribution: Madagascar

Angraecum minus Summerh.

Distribution: Ethiopia, Tanzania (United Republic of), Zambia, Zimbabwe

Angraecum minutissimum A.Chev.

Distribution: Cote d'Ivoire

Angraecum minutum Frapp. ex Cordem.
Angraecum minutum A.Chev.
Mystacidium minutum Cordem.

Distribution: Reunion

Angraecum mirabile Schltr.

Distribution: Madagascar

Angraecum moandense De Wild.
Angraecum chevalieri Summerh.
Aerangis moandensis (De Wild.) Schltr.

Distribution: Cameroon, Congo (The), Côte d'Ivoire, Equatorial Guinea, Gabon, Guinea, Liberia, Nigeria, Rwanda, Tanzania (United Republic of), Uganda

Angraecum modicum Summerh.

Distribution: Liberia

Angraecum mofakoko De Wild.

Distribution: Congo (The Democratic Republic of the)

Angraecum moratii Bosser

Distribution: Madagascar

Angraecum multiflorum Thouars
Aerobion multiflorum (Thouars) Spreng.
Angraecum caulescens var. *multiflorum* (Thouars) S.Moore
Epidorchis multiflora (Thouars) Kuntze
Monixus multiflorus (Thouars) Finet
Mystacidium caulescens ssp. *multiflorum* (Thouars) Durans & Schinz
Mystacidium multiflorum (Thouars) Cordem.

Distribution: Madagascar, Mauritius, Reunion, Seychelles

Angraecum multinominatum Rendle
Angraecum clavatum (Rendle) Schltr.
Listrostachys clavata Rendle
Macroplectrum clavatum (Rendle) Finet
Mystacidium clavatum (Rendle) Rolfe

Distribution: Gabon, Ghana, Guinea, Nigeria, Sierra Leone, Togo

Angraecum muscicolum H.Perrier

Distribution: Madagascar

Angraecum musculiferum H.Perrier

Distribution: Madagascar

Angraecum myrianthum Schltr.

Distribution: Madagascar

Angraecum nanum Frapp. ex Cordem.
Mystacidium nanum Cordem.

Distribution: Mauritius, Reunion

Angraecum nasutum Schltr.

Distribution: Madagascar

Angraecum nzoanum A.Chev.

Distribution: Guinea

Angraecum oberonia Finet

Distribution: Mauritius, Reunion

Angraecum obesum H.Perrier

Distribution: Madagascar

Angraecum oblongifolium Toill.-Gen. & Bosser

Distribution: Madagascar

Angraecum obversifolium Frapp.
Mystacidium obversifolium Cordem.

Distribution: Mauritius, Reunion

Angraecum ochraceum (Ridl.) Schltr.
Macroplectrum ochraceum (Ridl.) Finet
Mystacidium ochraceum Ridl.

Distribution: Madagascar

Angraecum onivense H.Perrier

Distribution: Madagascar

Angraecum palmicolum Bosser

Distribution: Madagascar

Angraecum palmiforme Thouars
Aerobion palmiforme Spreng.
Angorchis palmiformis Kuntze
Angraecum palmatum Thouars
Listrostachys palmiformis Durand & Schinz

Distribution: Mauritius, Reunion

Angraecum panicifolium H.Perrier

Distribution: Madagascar

Part II: Angraecum

Angraecum parvulum Ayres ex Baker
Angorchis parvula Kuntze

Distribution: Mauritius, Reunion

Angraecum pauciramosum Schltr.
Angraecum graminifolium (Ridl.) Schltr.
Angraecum poophyllum Summerh.
Epidorchis graminifolia (Ridl.) Kuntze
Monixus graminifolius (Ridl.) Finet
Mystacidium graminifolium Ridl.

Distribution: Madagascar

Angraecum pectinatum Thouars
Aeranthes pectinatus (Thouars) Rchb.f.
Aerobion pectinatum (Thouars) Spreng.
Angorchis pectangis (Thouars) Kuntze
Angorchis pectinata (Thouars) Kuntze
Ctenorchis pectinata (Thouars) K.Schum.
Macroplectrum pectinatum (Thouars) Finet
Mystacidium pectinatum (Thouars) Benth.
Pectinaria thouarsii Cordem.

Distribution: Comoros (The), Madagascar, Mauritius, Reunion

Angraecum penzigianum Schltr.

Distribution: Madagascar

Angraecum pergracile Schltr.

Distribution: Madagascar

Angraecum perhumile H.Perrier

Distribution: Madagascar

Angraecum perparvulum H.Perrier

Distribution: Madagascar

Angraecum petterssonianum Geerinck

Distribution: Rwanda

Angraecum peyrotii Bosser

Distribution: Madagascar

Angraecum pingue Frapp.
Mystacidium pingue Cordem.

Distribution: Mauritius, Reunion

Angraecum pinifolium Bosser

Distribution: Madagascar

Angraecum podochiloides Schltr.
Monixus aporum Finet

Distribution: Cameroon, Congo (The Democratic Republic of the), Côte d'Ivoire, Ghana, Liberia, Nigeria

Angraecum popowii Braem

Distribution: Madagascar

Angraecum potamophilum Schltr.
Aerangis potamophila (Schltr.) Schltr.

Distribution: Madagascar

Angraecum praestans Schltr.

Distribution: Madagascar

Angraecum protensum Schltr.

Distribution: Madagascar

Angraecum pseudodidieri H.Perrier

Distribution: Madagascar

Angraecum pseudofilicornu H.Perrier

Distribution: Madagascar

Part II: Angraecum

Angraecum pseudopetiolatum Frapp.
Mystacidium pseudo-petiolatum Cordem.

Distribution: Reunion

Angraecum pterophyllum H.Perrier

Distribution: Madagascar

Angraecum pumilio Schltr.

Distribution: Madagascar

Angraecum pungens Schltr.
Angraecum arthrophyllum (Kraenzl.) Schltr.
Mystacidium arthrophyllum Kraenzl.

Distribution: Cameroon, Congo (The Democratic Republic of the), Equatorial Guinea, Nigeria

Angraecum pusillum Lindl.
Angorchis pusilla (Lindl.) Kuntze
Angraecum burchellii Rchb.f.

Distribution: South Africa, Swaziland, Zimbabwe

Angraecum pygmaeum Linden

Distribution: Japan

Angraecum pyriforme Summerh.

Distribution: Côte d'Ivoire, Nigeria

Angraecum ramosum Thouars
Angraecum verruculosum Frapp. ex Cordem.

Distribution: Mauritius, Reunion

Angraecum ramulicolum H.Perrier

Distribution: Madagascar

Angraecum reygaertii De Wild.

Distribution: Cameroon, Congo (The Democratic Republic of the), Uganda

Angraecum rhizanthium H.Perrier

Distribution: Madagascar

Angraecum rhizomaniacum Schltr.

Distribution: Madagascar

Angraecum rhynchoglossum Schltr.
Angraecum foxii Summerh.
Angraecum viride (Ridl.) Schltr.
Epidorchis viridis (Ridl.) Kuntze
Mystacidium viride Ridl.

Distribution: Madagascar

Angraecum rigidifolium H.Perrier

Distribution: Madagascar

Angraecum rostratum Ridl.
Angorchis rostrata Kuntze

Distribution: Madagascar

Angraecum rubellum Bosser

Distribution: Madagascar

Angraecum rutenbergianum Kraenzl.
Angraecum catati Baill. ex H.Perrier
Jumellea rutenbergiana (Kraenzl.) Schltr.

Distribution: Madagascar

Angraecum sacciferum Lindl.
Angraecum crassifolia G.Will.
Angraecum parcum Schltr.

Distribution: Burundi, Cameroon, Congo (The Democratic Republic of the), Kenya, Malawi, Mozambique, Rwanda, South Africa, Swaziland, Tanzania (United Republic of), Uganda, Zambia, Zimbabwe

Angraecum sacculatum Schltr.

Distribution: Madagascar

Part II: Angraecum

Angraecum salazianum (Cordem.) Schltr.
Mystacidium salazianum Cordem.

Distribution: Reunion

Angraecum sambiranoense Schltr.

Distribution: Madagascar

Angraecum sanfordii P.J.Cribb & B.J.Pollard

Distribution: Cameroon

Angraecum scalariforme H.Perrier

Distribution: Madagascar

Angraecum scottianum Rchb.f.
Angorchis scottiana (Rchb.f.) Kuntze
Angraecum reichenbachianum Kraenzl.

Distribution: Comoros (The)

Angraecum sedifolium Schltr.

Distribution: Madagascar

Angraecum serpens (H.Perrier) Bosser
Jumellea serpens H.Perrier

Distribution: Madagascar

Angraecum sesquipedale Thouars
Aeranthes sesquipedalis (Thouars) Lindl.
Angorchis sesquipedalis (Thouars) Kuntze
Macroplectrum sesquipedale (Thouars) Pfitzer
Mystacidium sesquipedale (Thouars) Rolfe

Distribution: Madagascar

Angraecum sesquipedale var. **angustifolium** Bosser & Morat
Angraecum bosseri Senghas

Distribution: Madagascar

Angraecum sesquisectangulum Kraenzl.

Distribution: Madagascar

Angraecum setipes Schltr.

Distribution: Madagascar

Angraecum sinuatiflorum H.Perrier

Distribution: Madagascar

Angraecum sororium Schltr.

Distribution: Madagascar

Angraecum spectabile Summerh.

Distribution: Tanzania (United Republic of)

Angraecum spicatum (Cordem.) Schltr.
Mystacidium spicatum Cordem.

Distribution: Reunion

Angraecum stella-africae P.J.Cribb

Distribution: Malawi, South Africa, Zimbabwe

Angraecum sterrophyllum Schltr.
Angraecum filicornu sensu Kraenzl. non Thouars

Distribution: Madagascar

Angraecum stolzii Schltr.

Distribution: Congo (The Democratic Republic of the), Malawi, Mozambique, Tanzania (United Republic of), Zambia

Angraecum striatum Thouars
Aerobion striatum (Thouars) Spreng.
Angorchis striata (Thouars) Kuntze
Angraecum distichophyllum (A.Rich. ex Finet) Schltr.
Gastrochilus strictus Kuntze
Macroplectrum distichophyllum A.Rich. ex Finet
Monixus striatus (Thouars) Finet

Saccolabium striatum (Thouars) Lindl.

Distribution: Reunion

Angraecum subulatum Lindl.
Angraecum canaliculatum De Wild.
Epidorchis subulata (Lindl.) Kuntze
Listrostachys subulata (Lindl.) Rchb.f.

Distribution: Cameroon, Congo (The Democratic Republic of the), Côte d'Ivoire, Equatorial Guinea, Ghana, Nigeria, Sierra Leone

Angraecum tamarindicolum Schltr.

Distribution: Madagascar

Angraecum tenellum (Ridl.) Schltr.
Angraecum microphyton (Frapp.) Schltr.
Angraecum waterlotii H.Perrier
Epidorchis tenella (Ridl.) Kuntze
Mystacidium tenellum Ridl.
Saccolabium microphyton Frapp.

Distribution: Madagascar, Reunion

Angraecum tenuifolium Frapp. ex Cordem.
Lepervenchea tenuifolia Cordem.

Distribution: Reunion

Angraecum tenuipes Summerh.
Angraecum ischnopus Schltr.

Distribution: Madagascar

Angraecum tenuispica Schltr.

Distribution: Madagascar

Angraecum teres Summerh.

Distribution: Kenya, Tanzania (United Republic of)

Angraecum teretifolium Ridl.
Angorchis teretifolia (Ridl.) Kuntze
Monixus teretifolius (Ridl.) Finet

Distribution: Madagascar

Angraecum triangulifolium Senghas

Distribution: Madagascar

Angraecum trichoplectron (Rchb.f.) Schltr.
Aeranthes trichoplectron Rchb.f.
Mystacidium trichoplectron (Rchb.f.) T.Durand & Schinz

Distribution: Madagascar

Angraecum triquetrum Thouars
Aerobion triquetrum Spreng.
Angraecum neglectum Frapp. ex Cordem.
Angraecum neglectum ssp. *curtum* Cordem.
Angraecum neglectum ssp. *genuinum* Cordem.
Angraecum neglectum ssp. *longifolium* Cordem.

Distribution: Mauritius, Reunion

Angraecum umbrosum P.J.Cribb
Angraecum linearifolium P.J.Cribb non Garay

Distribution: Malawi

Angraecum undulatum (Cordem.) Schltr.
Mystacidium undulatum Cordem.

Distribution: Mauritius, Reunion

Angraecum urschianum Toill.-Gen. & Bosser

Distribution: Madagascar

Angraecum verecundum Schltr.

Distribution: Madagascar

Angraecum vesiculatum Schltr.

Distribution: Comoros (The), Madagascar

Angraecum vesiculiferum Schltr.

Distribution: Comoros (The), Madagascar

Angraecum viguieri Schltr.

Distribution: Madagascar

Angraecum viride Kraenzl.
Angraecum braunii Schltr.

Distribution: Kenya, Tanzania (United Republic of)

Angraecum viridiflorum Cordem.

Distribution: Reunion

Angraecum xylopus Rchb.f.
Macroplectrum xylopus (Rchb.f.) Finet

Distribution: Comoros (The)

Angraecum yuccaefolium Bojer

Distribution: Mauritius

Angraecum zaratananae Schltr.
Angraecum tsaratananae Schltr. ex H.Perrier

Distribution: Madagascar

Angraecum zeylanicum Lindl.
Angraecum maheense Schltr.

Distribution: Seychelles, Sri Lanka

ASCOCENTRUM BINOMIALS IN CURRENT USE

ASCOCENTRUM BINOMES ACTUELLEMENT EN USAGE

ASCOCENTRUM BINOMIALES UTILIZADOS NORMALMENTE

Ascocentrum ampullaceum (Roxb.) Schltr.
Aerides ampullacea Roxb.
Angraecum campyloplectron Rchb.f.
Gastrochilus ampullaceus (Roxb.) Kuntze
Saccolabium ampullaceum (Roxb.) Lindl.

Distribution: Bangladesh, Bhutan, China, India, Lao People's Democratic Republic (The), Myanmar, Nepal, Thailand, Viet Nam

Ascocentrum ampullaceum var. **auranticum** Pradhan

Distribution: Myanmar

Ascocentrum aurantiacum (Schltr.) Schltr.
Saccolabium aurantiacum Schltr. non J.J.Sm.

Distribution: Indonesia, Philippines (The)

Ascocentrum aurantiacum ssp. **philippinense** Christenson

Distribution: Philippines (The)

Ascocentrum aureum J.J.Sm.

Distribution: Indonesia

Ascocentrum christensonianum Haager

Distribution: Viet Nam

Ascocentrum curvifolium (Lindl.) Schltr.
Gastrochilus curvifolius (Lindl.) Kuntze
Saccolabium curvifolium Lindl.

Distribution: India, Myanmar, Thailand, Viet Nam

Ascocentrum garayi Christenson

Distribution: Cambodia, Lao People's Democratic Republic (The), Malaysia, Thailand, Viet Nam

Part II: Ascocentrum

Ascocentrum hendersoniana (Rchb.f.) Schltr.
Dyakia hendersoniana (Rchb.f.) Christenson
Saccolabium hendersonianum Rchb.f.

Distribution: Indonesia, Malaysia

Ascocentrum himalaicum (Deb., Sengupta & Malick) Christenson
Holcoglossum junceum Z.H.Tsi
Saccolabium himalaicum Deb., Sengupta & Malick

Distribution: Bhutan, China, India, Myanmar

Ascocentrum insularum Christenson

Distribution: Indonesia

Ascocentrum miniatum (Lindl.) Schltr.
Gastrochilus miniatus (Lindl.) Kuntze
Saccolabium curvifolium auct. non Lindl.
Saccolabium miniatum Lindl.

Distribution: Indonesia

Ascocentrum pumilum (Hayata) Schltr.
Ascolabium pumilum (Hayata) S.S.Ying
Saccolabium pumilum Hayata

Distribution: China

Ascocentrum pusillum Aver.
Ascocentropsis pussilum (Aver.) Senghas & Schildh.
Ascolabium pusillum (Aver.) Aver.

Distribution: Lao People's Democratic Republic (The), Thailand, Viet Nam

Ascocentrum rubrum (Lindl.) Seidenf.
Saccolabium rubrum Lindl.

Distribution: Myanmar

Ascocentrum semiteretifolium Seidenf.

Distribution: India, Thailand

BLETILLA BINOMIALS IN CURRENT USE

BLETILLA BINOMES ACTUELLEMENT EN USAGE

BLETILLA BINOMIALES UTILIZADOS NORMALMENTE

Bletilla chartacea (King & Pantl.) Tang & F.T.Wang
Bletilla burmanica Rolfe nom. nud.
Cephalanthera chartacea King & Pantl.

Distribution: Myanmar

Bletilla foliosa (King & Pantl.) Tang & F.T.Wang
Pogonia foliosa King & Pantl.

Distribution: Myanmar

Bletilla formosana (Hayata) Schltr.
Bletia formosana Hayata
Bletia kotoensis Hayata
Bletia morrisonicola Hayata
Bletilla cotoensis Schltr.
Bletilla elegantula (Kraenzl.) Garay & G.A.Romero
Bletilla formosana forma *kotoensis* (Hayata) T.P.Lin
Bletilla formosana forma *rubrolabella* S.S.Ying
Bletilla kotoensis (Hayata) Schltr.
Bletilla morrisonensis (Hayata) Schltr.
Bletilla morrisonicola (Hayata) Schltr.
Bletilla striata var. *kotoensis* (Hayata) Masam.
Bletilla szetschuanica Schltr. ex Limpr.
Bletilla yunnanensis Schltr. ex Limpr.
Bletilla yunnanensis var. *limprichtii* Schltr. ex Limpr.
Coelogyne elegantula Kraenzl.
Jimensia formosana (Hayata) Garay & R.E.Schultes
Jimensia kotoensis (Hayata) Garay & R.E.Schultes
Jimensia morrisonensis (Hayata) Garay & R.E.Schultes
Jimensia yunnanensis (Schltr.) Garay & R.E.Schultes

Distribution: China

Bletilla ochracea Schltr.
Jimensia ochracea (Schltr.) Garay & R.E.Schultes

Distribution: China, Viet Nam

Bletilla sinensis (Rolfe) Schltr.
Arethusa sinensis Rolfe
Bletilla chinensis Schltr.
Jimensia sinensis (Rolfe) Garay & R.E.Schultes

Distribution: China, Thailand

Part II: Bletilla

Bletilla striata (Thunb. ex A.Murray) Rchb.f.
Bletia gebina Lindl.
Bletia hyacinthina (Sm.) R.Br.
Bletia striata (Thunb. ex A.Murray) Druce
Bletilla gebina (Lindl.) Rchb.f.
Bletilla hyacinthina (Sm.) Rchb.f.
Bletilla striata var. *albomarginata* Mak.
Bletilla striata forma *gebina* (Lindl.) Rchb.f.
Calanthe gebina Lodd.
Cymbidium hyacinthinum Sm.
Cymbidium striatum (Thunb.) Sw.
Epidendrum striatum (Thunb.) Thunb.
Epidendrum tuberosum Lour.
Gyas humilis Salisb.
Jimensia nervosa Raf.
Jimensia striata (Thunb. ex A.Murray) Garay & R.E.Schultes
Limodorum striatum Thunb.

Distribution: China, Japan, Korea (The Democratic People's Republic of), Korea (The Republic of)

BRASSAVOLA BINOMIALS IN CURRENT USE

BRASSAVOLA BINOMES ACTUELLEMENT EN USAGE

BRASSAVOLA BINOMIALES UTILIZADOS NORMALMENTE

Brassavola acaulis Lindl. & Paxton
Bletia acaulis (Lindl.) Rchb.f.
Bletia lineata (Hook.) Rchb.f.
Brassavola lineata Hook.
Brassavola mathieuana Klotzsch

Distribution: Belize, Costa Rica, Guatemala, Mexico, Nicaragua, Panama

Brassavola angustata Lindl.
Bletia angustata (Lindl.) Rchb.f.
Bletia attenuata Rchb.f.
Brassavola surinamensis Focke

Distribution: Brazil, Guyana, Suriname, Venezuela

Brassavola cebolleta Rchb.f.
Bletia cebolleta Rchb.f.
Brassavola reginae Pabst
Brassavola revoluta sensu Withner

Distribution: Bolivia, Brazil

Brassavola chacoensis Kraenzl.
Brassavola cebolleta var. *fasciculata* (Pabst) H.G.Jones
Brassavola fasciculata Pabst
Brassavola ovaliformis C.Schweinf. sensu Withner
Brassavola ovaliformis var. *fasciculata* (Pabst) Jones sensu Withner

Distribution: Bolivia, Paraguay, Peru

Brassavola cordata Lindl.
Bletia cordata (Lindl.) Rchb.f.
Brassavola harrisii H.G.Jones
Brassavola nodosa Hook.
Brassavola nodosa var. *cordata* (Lindl.) N.H.Williams
Brassavola sloanei Griseb.
Brassavola subulifolia Lindl.
Lysimmia bicolor Rafin.

Distribution: Jamaica

Brassavola cucullata (L.) R.Br.
Bletia cucullata (L.) Rchb.f.
Brassavola appendiculata A.Rich. & Galeotti

Brassavola cucullata var. *elegans* Schltr.
Brassavola cuspidata Hook.
Brassavola elongata A.D.Hawkes *nomen nudum*
Brassavola odoratissima Regel
Cymbidium cucullatum (L.) Sw.
Epidendrum cucullatum L.

Distribution: Belize, Colombia, El Salvador, Guatemala, Guyana, Honduras, Mexico, Nicaragua, Trinidad and Tobago, United States of America (The), Venezuela

Brassavola filifolia Lind.

Distribution: Brazil

Brassavola flagellaris Barb.Rodr.

Distribution: Brazil

Brassavola gardneri Cogn.
Brassavola fragans Barb.Rodr.

Distribution: Brazil, Guyana, Suriname

Brassavola grandiflora Lindl.
Brassavola nodosa var. *grandiflora* (Lindl.) H.G.Jones

Distribution: Costa Rica, El Salvador, Guatemala, Honduras, Mexico, Nicaragua, Venezuela

Brassavola martiana Lindl.
Bletia amazonica Rchb.f.
Brassavola amazonica Poepp. & Endl.
Brassavola duckeana Horta
Brassavola martiana var. *multiflora* (Schltr.) H.G.Jones
Brassavola multiflora Schltr.
Brassavola paraensis Huber

Distribution: Bolivia, Brazil, Colombia, French Guiana, Guyana, Suriname, Venezuela

Brassavola nodosa (L.) Lindl.
Bletia nodosa (L.) Rchb.f.
Brassavola gillettei H.G.Jones
Brassavola nodosa var. *rhopalorrachis* (Rchb.f.) Schltr.
Brassavola rhopalorrachis Rchb.f.
Brassavola scaposa Schltr.
Brassavola sloanei Lindl. ex Heynh.
Brassavola stricta C.Kad.
Cymbidium nodosum Sw.

Epidendrum nodosum Linn.

Distribution: Belize, Brazil, Colombia, Costa Rica, Ecuador, El Salvador, Guatemala, Guyana, Honduras, Mexico, Nicaragua, Panama, ?Peru, Trinidad and Tobago,Venezuela

Brassavola perrinii Lindl.
Bletia perrinii (Lindl.) Rchb.f.
Brassavola perrinii var. *pluriflora* Hauman
Brassavola rhomboglossa Pabst

Distribution: Argentina, Bolivia, Brazil, Paraguay

Brassavola retusa Lindl.

Distribution: Brazil, Peru, Venezuela

Brassavola tuberculata Hook.
Brassavola fragrans Lem.
Brassavola gibbsiana Hort. ex Nichols
Brassavola revoluta Barb.Rodr.
Tulexis bicolor Raf.

Distribution: Brazil

Brassavola venosa Lindl.
Bletia venosa (Lindl.) Rchb.f.
Brassavola nodosa var. *venosa* (Lindl.) H.G.Jones

Distribution: Brazil, Colombia, Panama

CALANTHE BINOMIALS IN CURRENT USE

CALANTHE BINOMES ACTUELLEMENT EN USAGE

CALANTHE BINOMIALES UTILIZADOS NORMALMENTE

Calanthe abbreviata (Blume) Lindl.
Alismorchis abbreviata D.Kuntze
Amblyglottis abbreviata Blume

Distribution: Indonesia

Calanthe aceras Schltr.

Distribution: Papua New Guinea

Calanthe actinomorpha Fukuy.
Phaius actinomorphus (Fukuy.) T.P.Lin

Distribution: China

Calanthe × albolilacina J.J.Sm.

Distribution: Indonesia

Calanthe albo-longicalcarata S.S.Ying

Distribution: China

Calanthe albolutea Ridl.

Distribution: China, Indonesia, Malaysia

Calanthe aleizettei Gagnep.

Distribution: Viet Nam

Calanthe alismaefolia Lindl.
Alismorkis japonica (Blume ex Miq.) Kuntze
Calanthe austrokiusiuensis Ohwi.
Calanthe fauriei Schltr.
Calanthe furcata var. *alismifolia* (Lindl.) Hiroe
Calanthe furcata forma *faurie* (Lindl.) Hiroe
Calanthe japonica Blume ex Miq.
Calanthe nigropuncticulata Fukuy.

Distribution: Bhutan, China, India, Japan, Nepal, Viet Nam

Part II: Calanthe

Calanthe alpina Hook.f.
Calanthe alpina ssp. *fimbriata* (Fr.) F.Maek.
Calanthe alpina ssp. *fimbriatomarginata* (Fukuy.) F.Maek.
Calanthe alpina var. *keshabii* (Lucksom) R.C.Srivast.
Calanthe fimbriata Fr.
Calanthe fimbriatomarginata Fukuy.
Calanthe keshabii S.Z.Lucksom
Calanthe schlechteri Hara

Distribution: Bhutan, China, India, Japan, Nepal

Calanthe alta Rchb.f.
Calanthe anocentrum Schltr.
Calanthe lutescens Fleischm. & Rech.

Distribution: Fiji, Samoa

Calanthe angusta Lindl.
Alismorchis angusta (Lindl.) D.Kuntz.
Calanthe pachystalix Gagnep

Distribution: Cambodia, China, India, Lao People's Democratic Republic (The), Viet Nam

Calanthe angustifolia (Blume) Lindl.
Alismorkis angustifolia (Blume) Kuntze
Alismorkis phajoides (Rchb.f.) Kuntze
Amblyglottis angustifolia Blume
Calanthe phajoides Rchb.f.
Limodorum striatum Reinw. ex Hook.f.

Distribution: China, Indonesia, Malaysia, Philippines (The), Viet Nam

Calanthe anjanae Lucksom

Distribution: India

Calanthe anthropophora Ridl.

Distribution: Thailand

Calanthe arcuata Rolfe

Distribution: China

Calanthe arcuata var. **brevifolia** Z.H.Tsi

Distribution: China

Calanthe arfakana J.J.Sm.

Distribution: Indonesia

Calanthe argenteostriata C.Z.Tang & S.J.Cheng

Distribution: China, Viet Nam

Calanthe arisanensis Hayata
Calanthe sasakii Hayata

Distribution: China

Calanthe aristulifera Rchb.f.
Calanthe elliptica Hayata
Calanthe furcata forma *raishaensis* (Hayata) Hiroe
Calanthe kirishimensis Yatabe
Calanthe raishaensis Hayata

Distribution: China, Japan

Calanthe aruank P.Royen

Distribution: Papua New Guinea

Calanthe atjehensis J.J.Sm.

Distribution: Indonesia

Calanthe aurantiaca Ridl.

Distribution: Indonesia, Malaysia

Calanthe aurantimacula P. van Royen

Distribution: Indonesia, Papua New Guinea

Calanthe aureiflora J.J.Sm.

Distribution: Indonesia, Malaysia

Calanthe balansae Finet
Calanthe saccifera Kraenzl.

Distribution: New Caledonia (French)

Part II: Calanthe

Calanthe baliensis J.J.Wood & J.B.Comber

Distribution: Indonesia

Calanthe bicalcarata J.J.Sm.

Distribution: Indonesia

Calanthe biloba Lindl.
Calanthe biloba ssp. *obtusa* Rchb.f.

Distribution: China, India, Myanmar, Nepal, Thailand

Calanthe brachychila Gagnep.

Distribution: Viet Nam

Calanthe brevicornu Lindl.
Calanthe lamellosa Rolfe
Calanthe scaposa Z.H.Tsi & K.Y.Lang
Calanthe yunnanensis Rolfe

Distribution: Bhutan, China, India, Nepal

Calanthe breviflos Ridl.

Distribution: Indonesia

Calanthe buccinifera Rolfe ex Hemsl.

Distribution: China

Calanthe calanthoides (A.Rich. & Galeottii) Hamer & Garay
Calanthe cubensis Linden & Rchb.f.
Calanthe granatensis Rchb.f.
Calanthe mexicana Rchb.f.
Calanthe mexicana var. *retusa* Correll
Calanthe mexicana var. *lanceolata* Correll
Ghiesbreghtia calanthoides A.Rich. & Galeotti
Ghiesbreghtia mexicana A.Rich. & Galeotti ex Rchb.f.

Distribution: Colombia, Costa Rica, Cuba, Dominican Republic (the), Guatemala, Haiti, Honduras, Jamaica, Mexico, Panama

Calanthe camptoceras Schltr.

Distribution: Papua New Guinea

Calanthe candida Bosser

Distribution: Mauritius, Reunion

Calanthe cardioglossa Schltr.
Calanthe biloba auct. non Lindl.
Calanthe fuerstenbergiana Kraenzl. ex Schltr.
Calanthe hosseusiana Kraenzl.

Distribution: Lao People's Democratic Republic (The), Thailand, Viet Nam

Calanthe carrii Seidenf. & J.J.Wood
Calanthe pusilla Carr

Distribution: Malaysia

Calanthe caudatilabella Hayata
Calanthe caudatilabella var. *latiloba* Maek. ex Yamam.
Calanthe puberula var. *caudatilabella* (Hayata) Hiroe

Distribution: China

Calanthe caulescens J.J.Sm.
Calanthe apostasioides Schltr.
Calanthe salmoniviridis P.Royen

Distribution: Indonesia, Papua New Guinea

Calanthe caulodes J.J.Sm.

Distribution: Indonesia

Calanthe ceciliae Rchb.f.
Alismorchis zollingeri (Miq.) Kuntze
Amblyglottis veratrifolia Blume
Calanthe burmanica Rolfe
Calanthe scortechini Hook.f.
Calanthe sumatrana Blanco ex Boerl.
Calanthe wrayi Hook.f.
Calanthe zollingeri Miq. (non Rchb.f.)

Distribution: Indonesia, Malaysia, Myanmar, Thailand

Calanthe chevalieri Gagnep.

Distribution: Viet Nam

Part II: Calanthe

Calanthe chloroleuca Lindl.
Calanthe galeata Lindl.

Distribution: Bhutan, Nepal

Calanthe chrysoglossoides J.J.Sm.

Distribution: Indonesia

Calanthe chrysoleuca Schltr.

Distribution: Papua New Guinea

Calanthe clavata Lindl.
Calanthe clavata var. *malipoensis* Z.H.Tsi
Calanthe densiflora sensu King & Pantling non Lindl.

Distribution: China, India, Malaysia, Myanmar, Thailand, Viet Nam

Calanthe clavicalcar J.J.Sm.

Distribution: Indonesia

Calanthe cleistogama Holttum

Distribution: Malaysia

Calanthe coiloglossa Schltr.

Distribution: Papua New Guinea

Calanthe conspicua Lindl.
Calanthe lilacina Loher

Distribution: Malaysia, Philippines (The)

Calanthe coreana Nakai

Distribution: Korea (The Republic of)

Calanthe cremeoviridis J.J.Wood

Distribution: Papua New Guinea

Calanthe crenulata J.J.Sm.

Distribution: Indonesia, Malaysia

Calanthe cruciata Schltr.

Distribution: Papua New Guinea

Calanthe crumenata Ridl.

Distribution: Indonesia

Calanthe curvato-ascendens Gilli

Distribution: Papua New Guinea

Calanthe davaensis Ames

Distribution: Philippines (The)

Calanthe davidii Franch.
Calanthe bungoana Ohwi.
Calanthe davidii var. *bungoana* (Ohwi.) T.Hashim.
Calanthe ensifolia Rolfe
Calanthe furcata forma *matsudai* (Hayata) Hiroe
Calanthe matsudai Hayata

Distribution: China, Japan

Calanthe delavayi Finet
Calanthe coelogyniformis Kraenzl.

Distribution: China

Calanthe densiflora Lindl.
Calanthe kazuoi Yamam.
Phaius epiphyticus Seidenf.

Distribution: China, Japan, Myanmar, Nepal, Thailand, Viet Nam

Calanthe dipteryx Rchb.f.

Distribution: Indonesia

Calanthe discolor Lindl.
Calanthe cheniana Hand.-Mazz.

Part II: Calanthe

Calanthe esquirolii Schltr.

Distribution: China

Calanthe discolor forma **quinquelamellata** Hiroe

Distribution: Japan

Calanthe discolor ssp. **amamiana** (Fukuy.) Masam.
Calanthe amamiana Fukuy.
Calanthe amamiana ssp. *latilabella* Ida
Calanthe aristulifera var. *amamiana* (Fukuy.) Hatus.
Calanthe discolor forma *viridialba* (Maxim.) Honda
Calanthe discolor ssp. *latilabella* Ida
Calanthe discolor var. *amamiana* (Fukuy.) Masam.

Distribution: Japan

Calanthe discolor ssp. **discolor**
Alismographis lyroglossa Kuntze
Alismorkis discolor (Lindl.) Kuntze
Calanthe discolor ssp. *divaricatipetala* Ida
Calanthe discolor ssp. *viridialba* Maxim.
Calanthe lurida Decne.
Calanthe variegata Scheidw.

Distribution: China, Japan, Korea (The Democratic People's Republic of), Korea (The Republic of)

Calanthe discolor ssp. **kanashiroi** Fukuy.

Distribution: Japan

Calanthe discolor ssp. **tokunoshimensis** (Hatus. & Ida) Hatus.
Calanthe tokunoshimensis Hatus. & Ida
Calanthe tokunoshimensis forma *latilabella* Hats. & Ida

Distribution: Japan

Calanthe dulongensis H.Li

Distribution: China

Calanthe eberhardtii Gagnep.

Distribution: Viet Nam

Calanthe ecallosa J.J.Sm.

Distribution: Indonesia

Calanthe ecarinata Rolfe

Distribution: China

Calanthe emeishanica Z.H.Tsi & K.Y.Lang

Distribution: China

Calanthe engleriana Kraenzl.
Calanthe engleriana var. *brevicalcarata* J.J.Sm.
Calanthe latissimifolia R.S.Rogers

Distribution: Indonesia, Papua New Guinea

Calanthe epiphytica Carr

Distribution: Indonesia

Calanthe fargesii Finet

Distribution: China

Calanthe finisterrae Schltr.

Distribution: Papua New Guinea

Calanthe fissa L.O.Williams

Distribution: Papua New Guinea

Calanthe flava (Blume) Morren
Alismorchis parviflora Kuntze
Amblyglottis flava Blume
Calanthe veratrifolia Miq. non R.Br.
Calanthe parviflora Lindl.

Distribution: Indonesia

Calanthe forbesii Ridl.

Distribution: Indonesia

Part II: Calanthe

Calanthe formosana Rolfe
Calanthe disticha Tang & F.T.Wang
Calanthe patsinensis S.Y.Hu
Calanthe puberula var. *formosana* (Murata) Hiroe.
Calanthe pulchra var. *formosana* (Rolfe) S.S.Ying
Calanthe yushunii Mori & Yamam.

Distribution: China, Japan, Viet Nam

Calanthe fulgens Lindl.
Calanthe masuca var. *fulgens* (Lindl.) Hk.f.

Distribution: India

Calanthe geelvinkensis J.J.Sm.

Distribution: Indonesia

Calanthe gibbsiae Rolfe

Distribution: Indonesia, Malaysia

Calanthe graciliflora Hayata
Calanthe graciliflora var. *xuefengensis* Z.H.Tsi
Calanthe hamata Hand.-Mazz.

Distribution: China

Calanthe graciliscapa Schltr.

Distribution: Indonesia

Calanthe griffithii Lindl.
Calanthe plantaginea Griff. (non Lindl.)

Distribution: Bhutan, China, India, Myanmar

Calanthe halconensis Ames
Phaius calathoides Ames

Distribution: Philippines (The)

Calanthe hancockii Rolfe

Distribution: China

Calanthe hattorii Schltr.

Distribution: Japan

Calanthe hennisii Loher

Distribution: Philippines (The)

Calanthe henryi Rolfe

Distribution: China

Calanthe herbacea Lindl.

Distribution: China, India, Nepal, Viet Nam

Calanthe hirsuta Seidenf.

Distribution: China, Thailand

Calanthe hololeuca Rchb.f.
Calanthe neocaledonica Rendle
Calanthe neohibernica Schltr.
Calanthe vaupeliana Kraenzl.

Distribution: Fiji, New Caledonia, Papua New Guinea, Samoa, Solomon Islands, Tonga, Vanuatu

Calanthe hoshii S.Kobay.

Distribution: Japan

Calanthe humbertii H.Perrier

Distribution: Madagascar

Calanthe hyacinthina Schltr.

Distribution: Indonesia

Calanthe imthurnii Kores

Distribution: Fiji

Calanthe inflata Schltr.

Distribution: Papua New Guinea

Part II: Calanthe

Calanthe izu-insularis (Satomi) Ohwi. & Satomi
Calanthe aristulifera ssp. *izu-insularis* Satomi

Distribution: Japan

Calanthe johorensis Holttum

Distribution: Malaysia

Calanthe jusnerii Boxall ex Náves

Distribution: Philippines (The)

Calanthe kaniensis Schltr.

Distribution: Papua New Guinea

Calanthe kemulense J.J.Sm.

Distribution: Indonesia

Calanthe kinabaluensis Rolfe
Calanthe cuneata Ames & C.Schweinf.

Distribution: Malaysia

Calanthe labellicauda Gilli

Distribution: Papua New Guinea

Calanthe labrosa (Rchb.f) Rchb.f.
Calanthidium labrosum (Rchb.f.) Pfitzer
Limatodes labrosa Rchb.f.

Distribution: China, Myanmar, Thailand

Calanthe lacerata Ames

Distribution: Philippines (The)

Calanthe laxiflora Makino

Distribution: Japan

Calanthe lechangensis Z.H.Tsi

Distribution: China

Calanthe leucosceptrum Schltr.

Distribution: Indonesia, Papua New Guinea

Calanthe limprichtii Schltr.

Distribution: China

Calanthe longibracteata Ridl.

Distribution: Indonesia

Calanthe longifolia Schltr.

Distribution: Papua New Guinea, Soloman Islands

Calanthe lutiviridis P.Royen

Distribution: Papua New Guinea

Calanthe lyroglossa Rchb.f.
Alismorkis foerstermannii (Rchb.f.) Kuntze
Alismorkis lyroglossa (Rchb.f.) Kuntze
Calanthe foerstermannii Rchb.f.
Calanthe forsythiiflora Hayata
Calanthe liukiuensis Schltr.
Calanthe lyroglossa var. *forsythiiflora* (Hayata) S.S.Ying
Calanthe lyroglossa var. *longibracteata* P.O'Byrne
Calanthe nephroidea Gagnep.
Calanthe scortechinii Hook.f.

Distribution: Cambodia, China, Japan, Lao People's Democratic Republic (The), Malaysia, Myanmar, Philippines (The), Thailand, Viet Nam

Calanthe madagascariensis W.Watson ex Rolfe
Calanthe sylvatica sensu Rolfe
Calanthe warpuri W.Watson ex Rolfe

Distribution: Madagascar

Calanthe mannii Hook.f.
Calanthe oblanceolata Ohwi. & T.Koyama

Calanthe pusilla Finet

Distribution: Bhutan, China, India, Japan, Nepal, Viet Nam

Calanthe maquilingensis Ames

Distribution: Philippines (The)

Calanthe mcgregorii Ames

Distribution: Philippines (The)

Calanthe melinosema Schltr.

Distribution: Indonesia

Calanthe metoensis Z.H.Tsi

Distribution: China

Calanthe micrantha Schltr.
Calanthe fragrans P.Royen
Calanthe ombrophila P.Royen

Distribution: Papua New Guinea

Calanthe microglossa Ridl.

Distribution: Indonesia

Calanthe millotae Ursch & Toill.-Gen. ex Bosser

Distribution: Madagascar

Calanthe mindorensis Ames

Distribution: Philippines (The)

Calanthe moluccensis J.J.Sm.

Distribution: Indonesia

Calanthe monophylla Ridl.

Distribution: Malaysia

Calanthe nankunensis Z.H.Tsi

Distribution: China

Calanthe nephroglossa Schltr.

Distribution: Samoa

Calanthe nicolae P.O'Byrne

Distribution: Indonesia

Calanthe nipponica Mak.
Calanthe trulliformis ssp. *hastata* Finet

Distribution: China, Japan

Calanthe nivalis Boxall ex Náves

Distribution: Philippines (The)

Calanthe obreniformis J.J.Sm.

Distribution: Indonesia

Calanthe odora Griff.
Alismorkis angusta (Lindl.) Kuntze
Calanthe angusta Lindl.
Calanthe angusta var. *laeta* Hand.-Mazz.
Calanthe shweliensis W.W.Sm.
Calanthe vaginata Lindl.

Distribution: Bhutan, Cambodia, China, Nepal, Thailand, Viet Nam

Calanthe × **oodaruma** Maekawa

Distribution: Japan

Calanthe × **oreadum** Rendle

Distribution: New Caledonia (French)

Calanthe otuhanica C.L.Chan & T.J.Barkman

Distribution: Malaysia

Part II: Calanthe

Calanthe ovalifolia Ridl.

Distribution: Malaysia

Calanthe ovata Ridl.

Distribution: Malaysia

Calanthe pachystalix Rchb.f.

Distribution: Bhutan, India, Nepal

Calanthe parvilabris Schltr.

Distribution: Papua New Guinea

Calanthe pauciverrucosa J.J.Sm.

Distribution: Indonesia

Calanthe pavairiensis P.Ormerod

Distribution: Papua New Guinea, Solomon Islands

Calanthe petelotiana Gagnep.

Distribution: China, Viet Nam

Calanthe poilanei Gagnep.

Distribution: Cambodia, Lao People's Democratic Republic, Viet Nam

Calanthe plantaginea Lindl.
Calanthe plantaginea var. *lushuiensis* K.Y.Lang & Z.H.Tsi

Distribution: Bhutan, China, India, Nepal

Calanthe polyantha Gilli

Distribution: Papua New Guinea

Calanthe puberula Lindl.
Calanthe amoena W.W.Sm.
Calanthe lepida W.W.Sm.

Distribution: China, India, Nepal, Viet Nam

Calanthe pubescens Ridl.

Distribution: Malaysia

Calanthe pulchra (Blume) Lindl.
Alismorkis pulchra (Blume) Kuntze
Amblyglottis pulchra Blume
Calanthe curantigoides Kuntze
Calanthe curculigoides Lindl.
Styloglossum nervosum Breda

Distribution: India, Indonesia, Malaysia, Philippines (The), Thailand

Calanthe pullei J.J.Sm.

Distribution: Indonesia

Calanthe pumila Fukuy.
Calanthe striata var. *pumila* (Fukuy.) S.S.Ying

Distribution: China

Calanthe rajana J.J.Sm.

Distribution: Indonesia

Calanthe reflexa Maxim.
Alismorkis reflexa (Maxim.) Kuntze
Calanthe okushirensis Miyabe & Tatew.
Calanthe puberula var. *reflexa* (Maxim.) Hiroe
Calanthe puberula var. *okushirensis* (Miyabe & Tatewaki) Hiroe
Calanthe reflexa ssp. *okushirensis* (Miyabe & Tatew.) Ohwi.
Calanthe similis Schltr.
Paracalanthe reflexa (Maxim.) Kudo

Distribution: China, Japan, Korea (The Republic of)

Calanthe reflexilabris J.J.Sm.

Distribution: Indonesia

Calanthe repens Schltr.
Calanthe repens ssp. *pauliani* Ursch & Toill.-Gen.

Distribution: Madagascar

Calanthe rhodochila Schltr.
Calanthe bracteosa Schltr.

Calanthe breviscapa J.J.Sm.

Distribution: Indonesia, Papua New Guinea, Solomon Islands

Calanthe rhodochila var. **reconditiflora** (J.J.Sm.) S.Thomas
Calanthe reconditiflora J.J.Sm
Calanthe manis J.J.Sm.

Distribution: Indonesia, Papua New Guinea, Solomon Islands

Calanthe rigida Carr

Distribution: Indonesia, Malaysia

Calanthe rosea (Lindl.) Benth.
Alismorkis rosea (Lindl.) Kuntze
Limatodis rosea Lindl.

Distribution: Myanmar, Thailand

Calanthe rubens Ridl.
Calanthe elmeri Ames
Calanthe vestita var. *fournieri* sensu Seidenf.
Preptanthe rubens (Ridl.) Ridl.

Distribution: Malaysia, Philippines (The), Thailand, Viet Nam

Calanthe ruttenii J.J.Sm.

Distribution: Indonesia

Calanthe saccata J.J.Sm.

Distribution: Indonesia

Calanthe sacculata Schltr.
Calanthe sacculata var. *tchenkeoutinensis* Tang & F.T.Wang

Distribution: China

Calanthe salaccensis J.J.Sm.

Distribution: Indonesia, Malaysia

Calanthe seranica J.J.Sm.

Distribution: Indonesia

Calanthe shelfordii Ridl.

Distribution: Indonesia

Calanthe simplex Seidenf.

Distribution: China, Thailand

Calanthe sinica Z.H.Tsi

Distribution: China

Calanthe speciosa (Blume) Lindl.
Amblyglottis speciosa Blume
Calanthe speciosa Vieill.

Distribution: Indonesia, Malaysia

Calanthe stenophylla Schltr.

Distribution: Papua New Guinea

Calanthe striata R.Br. ex Lindl.
Calanthe bicolor Lindl.
Calanthe citrina Scheidw.
Calanthe discolor ssp. *bicolor* (Lindl.) Makino
Calanthe discolor ssp. *flava* Yatabe
Calanthe discolor forma *bicolor* (Lindl.) Hiroe
Calanthe kawakamii Hayata
Calanthe sieboldii Decne.
Calanthe striata Lindl.
Calanthe striata ssp. *bicolor* (Lindl.) Maxim
Calanthe striata var. *sieboldii* (Decne.) Maxim.
Calanthe takeoi Hayata
Limodorum striatum Banks (non Thunb.)

Distribution: China, Japan

Calanthe × subhamata J.J.Sm.

Distribution: Indonesia

Calanthe succedanea Gagnep.

Distribution: Cambodia, Lao People's Democratic Republic (The), Thailand, Viet Nam

Calanthe sylvatica (Thouars) Lindl.
Alimorchis masuca (D.Don) Kuntze

Part II: Calanthe

Alismorkis centrosis Steud.
Alismorkis pleiochroma (Rchb.f.) Kuntze
Alismorkis textori (Miq.) Kuntze
Amblyglottis emarginata Blume
Bletia masuca D.Don
Bletia silvatica (Thouars) Spreng.
Calanthe celebica Rolfe
Calanthe corymbosa Lindl.
Calanthe curtisii Rchb.f.
Calanthe delphinioides Kraenzl.
Calanthe durani Ursch & Toill.-Gen.
Calanthe emarginata (Blume) Lindl.
Calanthe furcata forma *masuca* (Lindl.) Hiroe
Calanthe furcata forma *textori* (Miq.) Hiroe
Calanthe kintaroi Yamam.
Calanthe longicalcarata Hayata ex Yamam.
Calanthe masuca (D.Don) Lindl.
Calanthe masuca forma *albiflora* (Ida) Nakajima
Calanthe masuca var. *purpurea* (Lindl.) Rao & Rathore
Calanthe masuca var. *sinensis* Rendle
Calanthe natalensis (Rchb.f.) Rchb.f.
Calanthe neglecta Schltr.
Calanthe okinawensis Hayata
Calanthe perrieri Ursch & Toill.-Gen.
Calanthe pleiochroma Rchb.f.
Calanthe purpurea Lindl.
Calanthe sanderiana B.S.Williams
Calanthe sanderiana Rolfe
Calanthe schliebenii Mansf.
Calanthe seikooensis Yamam.
Calanthe spathoglottoides Schltr.
Calanthe stolzii Schltr.
Calanthe sylvatica ssp. *natalensis* Rchb.f.
Calanthe sylvatica var. *pallidipetala* Schltr.
Calanthe sylvestris Lindl. ex Steud.
Calanthe textori Miq.
Calanthe textori forma *albiflora* Hatus.
Calanthe textori ssp. *alba* Maxim.
Calanthe textori ssp. *longicalcarata* (Hayata ex Yamam.) Garay & Sweet
Calanthe textori ssp. *textori*
Calanthe textori ssp. *violacea* Maxim.
Calanthe versicolor Lindl.
Calanthe violacea Rolfe
Calanthe volkensii Rolfe
Calanthe wightii Rchb.f.
Centrosia auberti nom. illeg. A.Rich.
Centrosis sylvatica Thouars
Zoduba masuca Buch.-Ham

Distribution: Angola, Bhutan, Burundi, Cameroon, China, Comoros (The), Congo (The Democratic Republic of the), Equatorial Guinea, Gabon, Guinea, India, Indonesia, Japan, Kenya, Madagascar, Malawi, Malaysia, Mauritius, Nepal, Papua New Guinea, Reunion, Rwanda, Sao Tome and Principe, Sierra Leone, South Africa, Sri Lanka, Swaziland, Tanzania (United Republic of), Thailand, Uganda, Zambia, Zimbabwe

Calanthe taenioides J.J.Sm.

Distribution: Indonesia, Malaysia

Calanthe tahitensis Nadeaud

Distribution: Tahiti

Calanthe tangmaiensis K.Y.Lang & Tateishi

Distribution: China

Calanthe tenuis Ames & C.Schweinf.
Calanthe fragilis R. Govaerts

Distribution: Malaysia

Calanthe torricellensis Schltr.

Distribution: Papua New Guinea

Calanthe transiens J.J.Sm.

Distribution: Indonesia, Malaysia

Calanthe triantherifera Nadeaud

Distribution: French Polynesia

Calanthe tricarinata Lindl.
Calanthe lamellata Hayata
Calanthe megalopha Fr.
Calanthe occidentalis Lindl.
Calanthe pantlingii Schltr.
Calanthe torifera Schltr.
Calanthe undulata Schltr.
Paracalanthe tricarinata (Lindl.) Kudo

Distribution: Bhutan, China, India, Japan, Nepal, Thailand

Calanthe trifida Tang & F.T.Wang

Distribution: China

Calanthe triplicata (Willem.) Ames
Alismorkis diploxiphion (Hook.f.) Kuntze
Alismorkis furcata (Bateman) Kuntze
Alismorkis gracillima (Lindl.) Kuntze

Alismorkis veratrifolia (Willd.) Kuntze
Amblyglottis veratrifolia (Willd.) Blume
Calanthe anchorifera Rchb.f.
Calanthe angraeciflora Rchb.f.
Calanthe bracteosa Rchb.f.
Calanthe brevicolumna Hayata
Calanthe catilligera Rchb.f.
Calanthe comosa Rchb.f.
Calanthe diploxiphion Hook.f.
Calanthe elytroglossa Rchb.f. ex Hook.f.
Calanthe furcata Bateman ex Lindl.
Calanthe furcata forma *albo-lineata* Nakajima
Calanthe furcata forma *albo-marginata* Nakajima
Calanthe furcata forma *brevicolumna* Hayata
Calanthe gracillima Lindl.
Calanthe matsumurana Schltr.
Calanthe muelleri Kraenzl.
Calanthe orthocentron Schltr.
Calanthe perrottetii A.Rich.
Calanthe petri Rchb.f.
Calanthe proboscidea Rchb.f.
Calanthe rubicallosa Masam.
Calanthe triplicata forma *albo-lineata* (Nakajima) Hatsu.
Calanthe triplicata forma *albo-marginata* (Nakajima) Nakajima
Calanthe triplicata var. *angraeciflora* (Rchb.f.) N.Halle
Calanthe triplicata var. *gracillima* (Lindl.) N.Halle
Calanthe veratrifolia (Willd.) Ker Gawl.
Calanthe veratrifolia (Willd.) R.Br.
Calanthe veratrifolia ssp. *densissima* J.J.Sm.
Calanthe veratrifolia ssp. *dupliciloba* J.J.Sm.
Calanthe veratrifolia ssp. *incurvicalca* J.J.Sm.
Calanthe veratrifolia ssp. *incurvicalcar* J.J.Sm.
Calanthe veratrifolia ssp. *lancipetala* J.J.Sm.
Calanthe veratrifolia ssp. *timorensis* J.J.Sm.
Calanthe veratrifolia var. *australis* Lindl. & Paxton
Calanthe veratrifolia var. *kennyi* F.M. Bailey
Calanthe veratrifolia var. *stenochila* Rchb.f.
Flos triplicatus Rumph.
Limodorum ventricosum Steud.
Limodorum veratrifolium Willd.
Orchis triplicata Willem.

Distribution: Australia, Bhutan, Cambodia, China, ?Fiji, India, Indonesia, Japan, Lao People's Democratic Republic (The), Malaysia, Mauritius, Micronesia (Federated States of), Myanmar, Nepal, New Caledonia, Papua New Guinea, Philippines (The), Samoa, Solomon Islands, Sri Lanka, Thailand, Viet Nam

Calanthe truncata J.J.Sm.

Distribution: Indonesia

Calanthe truncicola Schltr.

Distribution: Indonesia, Malaysia

Calanthe tsoongiana Tang & F.T.Wang
Calanthe tsoongiana var. *guizhouensis* Z.H.Tsi

Distribution: China

Calanthe tyoh-harai Makino

Distribution: Japan

Calanthe uncata Lindl.

Distribution: India

Calanthe undulata J.J.Sm.

Distribution: Indonesia, Malaysia

Calanthe unifolia Ridl.

Distribution: Indonesia

Calanthe × varians J.J.Sm.

Distribution: Indonesia

Calanthe velutina Ridl.
Calanthe integrilabris Gagnep.

Distribution: Viet Nam

Calanthe ventilabrum Rchb.f.
Calanthe bigibba Schltr.
Calanthe chrysantha Schltr.
Calanthe englishii Rolfe
Calanthe flabelliformis Rogers
Calanthe langei F.Muell.
Calanthe tunensis J.J.Sm.

Distribution: Fiji, Indonesia, New Caledonia, Papua New Guinea, Samoa, Solomon Islands, Vanuatu

Calanthe versteegii J.J.Sm.
Calanthe arundinoides J.J.Sm

Distribution: Indonesia

Calanthe vestita Lindl.
Amblyglottis pilosa de Vriese ex Lindl.
Aulostylis papuana Schltr.
Calanthe augusti-reigneri Rchb.f.
Calanthe grandiflora Rolfe
Calanthe inquilinax Rolfe
Calanthe padangensis Schltr. ex Mansf.
Calanthe papuana (Schltr.) J.J.Sm.
Calanthe pilosa (de Vriese) Miq.
Calanthe regnieri Rchb.f.
Calanthe stevensiana Regnier ex Rchb.f.
Calanthe stevensii Hort. ex Will.
Calanthe turneri Rchb.f.
Calanthe vestita ssp. *fournieri* Rolfe
Calanthe vestita ssp. *igneo-oculata* Hort. ex Rchb.f.
Calanthe vestita ssp. *oculata-gigantea* Rchb.f. ex Williams
Cytheris griffithii Wight
Phaius vestitus (Lindl.) Rchb.f.
Preptanthe vestita (Lindl.) Rchb.f.
Preptanthe villosa Rchb.f.

Distribution: Indonesia, Malaysia, Myanmar, Papua New Guinea, Philippines (The), Thailand, Viet Nam

Calanthe villosa J.J.Sm.

Distribution: Indonesia

Calanthe whiteana King & Pantl.
Calanthe wardii W.W.Sm.

Distribution: China, India

Calanthe yoksomnensis Lucksom

Distribution: India

Calanthe yuana Tang & F.T.Wang

Distribution: China

Calanthe zollingeri Rchb.f.
Calanthe mutabilis Ridl.
Calanthe zollingeri var. *longecalcarata* J.J.Sm.

Distribution: Indonesia

CATASETUM BINOMIALS IN CURRENT USE

CATASETUM BINOMES ACTUELLEMENT EN USAGE

CATASETUM BINOMIALES UTILIZADOS NORMALMENTE

Catasetum aculeatum F.E.L.Miranda & K.G.Lacerda

Distribution: Brazil

Catasetum adremedium D.E.Benn. & Christenson

Distribution: Peru

Catasetum albovirens Barb.Rodr.

Distribution: Brazil

Catasetum aripuanense Bicalho

Distribution: Brazil

Catasetum atratum Lindl.
Catasetum adnatum Steud.
Catasetum atratum var. *mentosum* Mansf.
Catasetum mentosum Lem.
Catasetum pallidum Klotzsch

Distribution: Brazil

Catasetum barbatum (Lindl.) Lindl.
Catasetum appendiculatum Schltr. sensu Dunst.
Catasetum barbatum var. *probiscidium* Lindl.
Catasetum barbatum var. *spinosum* Rolfe
Catasetum brachybulbon Schltr. sensu Dunst.
Catasetum buchtienii Kraenzl. sensu Dunst.
Catasetum comosum Cogn.
Catasetum craniomorphum Hoffmgg. ex Heynh.
Catasetum crashleyanum St. Leger
Catasetum crinitum Linden
Catasetum cristatum var. *spinigerum* Hook.
Catasetum cristatum var. *spinosum* Hook.
Catasetum garnettianum Rolfe sensu Dust.
Catasetum macrocarpum Stein non Rich. ex Kunth
Catasetum polydactylon Schltr. sensu Dunst.
Catasetum proboscideum Lindl.
Catasetum randii Rolfe sensu Dunst.
Catasetum rivularium Barb.Rodr. sensu Dunst.
Catasetum spinosum Lindl. sensu Dunst.
Catasetum tridentatum Pfitzer non Hook.

Part II: Catasetum

Catasetum variabile Rodrig.
Myanthus barbatus Lindl.
Myanthus spinosus Hook. sensu Dunst.

Distribution: Brazil, Colombia, Ecuador, Guyana, Peru, Venezuela

Catasetum bergoldianum Foldats

Distribution: Brazil, Colombia, Venezuela

Catasetum bicallosum Cogn.
Catasetum stenoglossum Pabst

Distribution: Colombia, Venezuela

Catasetum bicolor Klotzsch
Catasetum gongoroides Kraenzl.

Distribution: Colombia, Panama, Venezuela

Catasetum blackii Pabst

Distribution: Brazil

Catasetum boyi Mansf.

Distribution: Brazil

Catasetum brachybulbon Schltr.

Distribution: Brazil

Catasetum buchtienii Kraenzl.

Distribution: Bolivia

Catasetum callosum Lindl.
Catasetum acallosum Lindl. ex Rchb.f.
Catasetum arachnoides Ames
Catasetum callosum var. *carunculatum* Mansf.
Catasetum callosum var. *crenatum* Regel
Catasetum callosum var. *eucallosum* Mansf.
Catasetum callosum var. *grandiflorum* Hook. in Curtis
Catasetum callosum var. *typum* Hoehne
Catasetum carunculatum Rchb.f. & Warz.
Catasetum darwinianum Rolfe
Catasetum fuliginosum Rolfe non Lindl.
Catasetum landsbergii (Reinw. & de Vriese) Lindl. & Paxton

Myanthus callosus (Lindl.) Beer
Myanthus grandiflorum Beer
Myanthus landsbergii Reinw. & de Vriese

Distribution: Brazil, Colombia, Peru, Venezuela

Catasetum caputinum da Silva

Distribution: Brazil

Catasetum carolinianum F.E.Miranda & K.G.Lacerda

Distribution: Brazil

Catasetum cernuum (Lindl.) Rchb.f.
Catasetum cernuum var. *cernuum* Mansf.
Catasetum cernuum var. *revolutum* Cogn.
Catasetum cernuum var. *rodigasianum* Mansf.
Catasetum cernuum var. *typum* Hoehne
Catasetum cernuum var. *umbrosum* Cogn.
Catasetum cernuum var. *umbrosum* (Barb.Rodr.) Cogn.
Catasetum rodigasianum Rolfe
Catasetum rohrii Pabst
Catasetum trifidum Hook.
Catasetum umbrosum Barb.Rodr.
Catasetum viride Lindl.
Monachanthus viridis (Lindl.) Rchb.f.
Myanthus cernuus Lindl.

Distribution: Brazil, Trinidad and Tobago

Catasetum charlesworthii (Mansfield) R. Jenny
Catasetum naso var. *charlesworthii* Mansf.

Distribution: Colombia

Catasetum cochabambanum Dodson & Vasquez

Distribution: Bolivia

Catasetum collare Cogn.

Distribution: Brazil, Colombia, Venezuela

Catasetum complanatum F.E.Miranda & K.G.Lacerda

Distribution: Brazil

Part II: Catasetum

Catasetum confusum G.A.Romero

Distribution: Brazil

Catasetum coniforme C.Schweinf.

Distribution: Peru

Catasetum cornutum Lindl.
Catasetum bicornutum Hort. ex Planchon

Distribution: Guyana

Catasetum costatum Rchb.f.

Distribution: Brazil, Venezuela

Catasetum cotylicheilum D.E.Benn. & Christenson

Distribution: Peru

Catasetum cristatum Lindl.
Catasetum cristatum var. *monstruosum* Hook.
Catasetum cristatum var. *stenosepalum* Rchb.f.
Catasetum cristatum var. *supralobatum* Branger
Monachanthus cristatus (Lindl.) Lindl.
Myanthus cristatus (Lindl.) Lindl.

Distribution: Brazil

Catasetum cucullatum A.T.Oliveira & da Silva

Distribution: Brazil

Catasetum decipiens Rchb.f.

Distribution: Venezuela

Catasetum deltoideum (Lindl.) Mutel
Myanthus deltoideus Lindl.

Distribution: Brazil

Catasetum denticulatum Miranda

Distribution: Brazil

Catasetum discolor (Lindl.) Lindl.
Catasetum cassideum Linden & Rchb.f.
Catasetum claesianum L.Lind & Cogn.
Catasetum discolor forma *genuinum* Hoehne
Catasetum discolor var. *bushnani* Cogn.
Catasetum discolor var. *claesianum* (L. Linden & Cogn.) Mansf.
Catasetum discolor var. *claesianum* Mansf.
Catasetum discolor var. *discolor* Mansf.
Catasetum discolor var. *viridiflorum* Cogn.
Catasetum gardneri Schltr.
Monachanthus bushnani Hook.
Monachanthus discolor Lindl.
Monachanthus discolor var. *bushnani* Hook.
Monachanthus discolor var. *viridiflorus* Hook.

Distribution: Brazil, Colombia, Venezuela

Catasetum × dunstervillei G.A.Romero & Carnevali

Distribution: Brazil, Venezuela

Catasetum dupliciscululatum Senghas

Distribution: Bolivia

Catasetum expansum Rchb.f.
Catasetum cliftonii Hort
Catasetum platyglossum Schltr.

Distribution: Ecuador

Catasetum faustii Hort. ex da Silva

Distribution: Brazil

Catasetum fernandezii D.E.Benn. & Christenson

Distribution: Peru

Catasetum ferox Kraenzl.
Catasetum brichtiae Bicalho

Distribution: Brazil

Catasetum fimbriatum (Morren) Lindl. & Paxton
Catasetum cogniauxii L.Linden
Catasetum fimbriatum var. *aurantiacum* Porsch
Catasetum fimbriatum var. *brevipetalum* Porsch
Catasetum fimbriatum var. *cogniauxii* L.

Catasetum fimbriatum var. *fissum* Rchb.f.
Catasetum fimbriatum var. *inconstans* Mansf.
Catasetum fimbriatum var. *micranthum* Porsch
Catasetum fimbriatum var. *morrenianum* Mansf.
Catasetum fimbriatum var. *ornithorrhynchum* Mansf.
Catasetum fimbriatum var. *platypterum* Rchb.f.
Catasetum fimbriatum var. *subtropicale* Hauman
Catasetum fimbriatum var. *viridulum* Rchb.f.
Catasetum inconstans Hoehne
Catasetum negrense Schltr.
Catasetum ornithorhynchum Porsch
Catasetum pflanzii Schltr.
Catasetum wredeanum Schltr.
Myanthus fimbriatus Morren

Distribution: Bolivia, Brazil, Colombia, Ecuador, Venezuela

Catasetum franchinianum K.G.Lacerda

Distribution: Brazil

Catasetum fuchsii Dodson & Vasquez

Distribution: Bolivia

Catasetum galeatum K.G.Lacerda

Distribution: Brazil

Catasetum galeritum Rchb.f.
Catasetum galeritum var. *galeritum* Mansf.
Catasetum galeritum var. *pachyglossum* Rchb.f.

Distribution: Brazil

Catasetum garnettianum Rolfe

Distribution: Brazil

Catasetum georgii Mansf.
Catasetum huebneri Mansf. non Schltr.

Distribution: Brazil

Catasetum gladiatorium K.G.Lacerda
Catasetum appendiculatum Schltr.

Distribution: Brazil

Catasetum globiflorum Hook.
Catasetum globiferum Beer (sphalm)

Distribution: Brazil

Catasetum gnomus Lind. & Rchb.f.
Catasetum gnomus var. *phasma* (Rchb.f.) Cogn.
Catasetum heteranthum Barb.Rodr.
Catasetum huebneri sensu Romero & Jenny
Catasetum phasma Rchb.f.

Distribution: Brazil

Catasetum gomezii G.A.Romero & Carnevali

Distribution: Venezuela

Catasetum × guianense G.A.Romero & Jenny

Distribution: Guyana, Suriname

Catasetum hillsii D.E.Benn. & Christenson

Distribution: Peru

Catasetum hookeri Lindl.
Catasetum hookeri var. *labiatum* (Barb.Rodr.) Cogn.
Catasetum hookeri var. *triste* (Rchb.f.) Rchb.f.
Catasetum imschootianum L.Lind & Cogn.
Catasetum labiatum Barb.Rodr.
Catasetum milleri Lodd. ex Lindl.
Catasetum triste Rchb.f.

Distribution: Brazil

Catasetum huebneri Schltr.

Distribution: Brazil, Venezuela

Catasetum imperiale L.Lind & Cogn.

Distribution: Brazil

Catasetum incurvum Klotzsch
Catasetum stupendum Cogn.
Catasetum trautmannii Senghas

Distribution: Peru

Part II: Catasetum

Catasetum integerrimum Hook.
Catasetum integerrimum var. *purpurascens* Hook.
Catasetum integerrimum var. *viridiflorum* Hook.
Catasetum integrinum Beer (sphalm)
Catasetum maculatum auct. non Kunth
Catasetum wailesii Hook.

Distribution: Belize, Costa Rica, El Salvador, Guatemala, Honduras, Mexico, Nicaragua

Catasetum × intermedium L.C.Menezes & Braem
Catasetum intermedium var. *rubrum* L.C.Menezes
Catasetum intermedium var. *zebrinum* L.C.Menezes

Distribution: Brazil

Catasetum × issanensis Pabst

Distribution: Brazil

Catasetum juruense Hoehne

Distribution: Brazil

Catasetum justinianum Vasquez & Dodson

Distribution: Bolivia

Catasetum kempfii Dodson & Vasquez

Distribution: Bolivia

Catasetum kleberianum Braga

Distribution: Brazil

Catasetum kraenzlinianum Mansf.
Catasetum micranthum Kraenzl. non Barb.Rodr.

Distribution: Brazil

Catasetum laminatum Lindl.
Catasetum lamilatum Beer (sphalm)
Catasetum laminatum var. *eburneum* Lindl.
Catasetum laminatum var. *maculatum* Lindl.

Distribution: Mexico

Catasetum lanceanum Miranda

Distribution: Brazil

Catasetum lanciferum Lindl.
Catasetum appendiculatum Schltr. sensu K.G.Lacerda

Distribution: Brazil

Catasetum lanxiforme Senghas

Distribution: Peru

Catasetum lemosii Rolfe
Catasetum roseum Barb.Rodr. non (Lindl.) Rchb.f.

Distribution: Brazil

Catasetum lindleyanum Mansf.

Distribution: ?Colombia, ?Ecuador

Catasetum linguiferum Schltr.

Distribution: Brazil

Catasetum longifolium Lindl.
Monachanthus longifolius (Lindl.) Hook.

Distribution: Brazil, Colombia, French Guiana, Guyana, Suriname, Venezuela

Catasetum longipes F.E.Miranda & K.G.Lacerda

Distribution: Brazil

Catasetum lucis P.Ortiz & Arango
Catasetum lucis forma *tigrinum* P.Ortiz. & Arango

Distribution: Colombia

Catasetum luridum (Link) Lindl.
Anguloa lurida Link
Catachaetum craniomorphum Hoffmgg.
Catachaetum lituratum Hoffmgg.
Catachaetum purpurascens Hoffmgg.
Catachaetum squalidum Hoffmgg.
Catachaetum turbinatum Hoffmgg.

Catasetum abruptum Hook.
Catasetum lituratum Hoffmgg.
Catasetum purpurascens Hoffmgg. ex Heynh.
Epidendrum ollare Vell.

Distribution: Brazil

Catasetum macrocarpum Rich. ex Kunth
Catasetum claveringii Lindl.
Catasetum floribundum Hook.
Catasetum grandis Hort ex Planchon
Catasetum integerrimum var. *flavescens* (Cogn.) Mansf.
Catasetum macrocarpum var. *amplissimum* (Planchon) Cogn.
Catasetum macrocarpum var. *bellum* Rchb.f.
Catasetum macrocarpum var. *brevifolium* (Mutel) Cogn.
Catasetum macrocarpum var. *carnosissimum* Cogn.
Catasetum macrocarpum var. *chrysanthum* Lind & Rod
Catasetum macrocarpum var. *genuinum* (Mutel) Cogn.
Catasetum macrocarpum var. *globoso-connivens* (Mutel) Cogn.
Catasetum macrocarpum var. *lindeni* O'Brien
Catasetum macrocarpum var. *pallidum* (Mutel) Cogn.
Catasetum macrocarpum var. *unidentatum* (Mutel) Cogn.
Catasetum macrocarpum var. *viridi-eburneum* (Mutel) Cogn.
Catasetum macrocarpum var. *viridi-sanguineum* (Mutel) Cogn.
Catasetum maculatum var. *flavescens* Cogn.
Catasetum menthaeodorum Hort. ex Planchon
Catasetum tricolor Hort. ex Planchon
Catasetum tridentatum Hook.
Catasetum tridentatum var. *amplissimum* Planchon
Catasetum tridentatum var. *breviflorum* Mutel
Catasetum tridentatum var. *claveringii* Planchon
Catasetum tridentatum var. *floribundrum* (Hook.) Planchon
Catasetum tridentatum var. *geniunum* Mutel
Catasetum tridentatum var. *globoso-connivens* Mutel
Catasetum tridentatum var. *macrocarpum* Planchon
Catasetum tridentatum var. *maximum* Planchon
Catasetum tridentatum var. *pallidum* Mutel
Catasetum tridentatum var. *unidentatum* Mutel
Catasetum tridentatum var. *viridi-eburnem* Mutel
Catasetum tridentatum var. *viridiflorum* Hook. in Curtis
Catasetum tridentatum var. *viridi-sanguinem* Mutel

Distribution: Brazil, Colombia, French Guiana, Guyana, Suriname, Venezuela

Catasetum macroglossum Rchb.f.

Distribution: Ecuador

Catasetum maculatum Kunth
Catasetum blepharochilum Schltr.
Catasetum brenesii Schltr.
Catasetum oerstedii Rchb.f.

Catasetum rostratum Klinge

Distribution: Costa Rica, Nicaragua, Panama, Venezuela

Catasetum maranhense K.G.Lacerda & De Silva

Distribution: Brazil

Catasetum maroaense G.A.Romero & C.Gomez

Distribution: Venezuela

Catasetum mattogrossense Bicalho

Distribution: Brazil

Catasetum mattosianum Bicalho

Distribution: Brazil

Catasetum meeae Pabst

Distribution: Brazil

Catasetum merchae G.A.Romero

Distribution: Venezuela

Catasetum micranthum Barb.Rodr.

Distribution: Brazil

Catasetum microglossum Rolfe

Distribution: Colombia, Ecuador, Peru, Venezuela

Catasetum mocuranum Schltr.

Distribution: Brazil

Catasetum mojuense A.T.Oliveira & da Silva

Distribution: Brazil

Catasetum monzonensis D.E.Benn. & Christenson

Distribution: Peru

Catasetum moorei C.Schweinf.

Distribution: Peru

Catasetum multifidum Miranda

Distribution: Brazil

Catasetum multifissum Senghas

Distribution: Peru

Catasetum nanayanum Dodson & D.E.Benn.

Distribution: Peru

Catasetum napoense Dodson

Distribution: Ecuador

Catasetum naso Lindl.
Catasetum naso var. *naso* Mansf.

Distribution: Colombia, Venezuela

Catasetum ochraceum Lindl.
Catasetum inornatum Schltr.
Catasetum lehmannii Regel

Distribution: Colombia

Catasetum ollare L.Linden

Distribution: Brazil

Catasetum ornithoides Pabst

Distribution: Brazil

Catasetum osculatum K.G.Lacerda & V.P.Castro

Distribution: Brazil

Catasetum parguazense G.A.Romero & Carnevali

Distribution: Brazil, Venezuela

Catasetum pendulum Dodson

Distribution: Mexico

Catasetum peruvianum Dodson & D.E.Benn.

Distribution: Peru

Catasetum pileatum Rchb.f.
Catasetum bungerothi N.E.Br.
Catasetum bungerothi var. *album* Linden & Rodigas
Catasetum bungerothi var. *aurantiacum* Cogn.
Catasetum bungerothi var. *lindeni* Cogn.
Catasetum bungerothi var. *pottsianum* Linden & Rodig
Catasetum bungerothi var. *randi* Rodigas
Catasetum bungerothi var. *regale* Cogn.
Catasetum bungerothii var. *imperiale* Cogn.
Catasetum pileatum var. *album* (Linden & Rodigas) Hoehne
Catasetum pileatum var. *aurantiacum* (Cogn.) Hoehne
Catasetum pileatum var. *imperiale* (Linden & Cogn.) Cogn.
Catasetum pileatum var. *lindeni* (Gower) Hoehne
Catasetum pileatum var. *pottisianum* (Linden & Rodigas) Hoehne
Catasetum pileatum var. *randii* (Rodigas) Hoehne
Catasetum pileatum var. *regale* (Gower) Hoehne

Distribution: Brazil, Colombia, Venezuela

Catasetum planiceps Lindl.
Catachaetum recurvatum (Lindl.) Hoffmans.
Catasetum hymenophorum Cogn.
Catasetum recurvatum Link, Klotzsch & Otto

Distribution: Brazil, Colombia, Guyana, Panama, Venezuela

Catasetum pleiodactylon D.E.Benn. & Christenson

Distribution: Peru

Catasetum × pohlianum G.Castro & Campacci

Distribution: Brazil

Catasetum polydactylon Schltr.

Distribution: Brazil

Part II: Catasetum

Catasetum poriferum Lindl.

Distribution: Guyana, Venezuela

Catasetum pulchrum N.E.Br.
Catasetum cirrhaeoides Hoehne
Catasetum cirrhaeoides var. *hoehnei* Mansf.
Catasetum cirrhaeoides var. *longicirrhosa* Mansf.
Catasetum jarae Dodson & D.E.Benn.

Distribution: Brazil, Peru

Catasetum punctatum Rolfe

Distribution: Brazil

Catasetum purum Nees & Sinning
Catachaetum purum (Nees & Sinning) Hoffmanns.
Catachaetum semiapertum (Hook.) Hoffmanns.
Catasetum immaculatum Hort ex Planchon
Catasetum inapertum Steud.
Catasetum semiapertum Hook.

Distribution: Brazil

Catasetum purusense D.E.Benn. & Christenson

Distribution: Peru

Catasetum pusillum C.Schweinf.

Distribution: Peru

Catasetum quadridens Rolfe

Distribution: Brazil

Catasetum randii Rolfe

Distribution: Brazil

Catasetum regnellii Barb.Rodr.

Distribution: Brazil

Catasetum reichenbachianum Mansf.

Distribution: Brazil

Catasetum richteri Bicalho

Distribution: Brazil

Catasetum ricii Vasquez & Dodson

Distribution: Bolivia

Catasetum rivularium Barb.Rodr.

Distribution: Brazil, Venezuela

Catasetum rolfeanum Mansf.
Catasetum stenochilum Kraenzl.

Distribution: Brazil

Catasetum rondonense Pabst

Distribution: Brazil

Catasetum rooseveltianum Hoehne

Distribution: Bolivia, Brazil

Catasetum × **roseo-album** (Hook.) Lindl.
Catasetum ciliatum Barb.Rodr.
Catasetum discolor var. *roseo-album* Mansf.
Catasetum discolor var. *vinosum* Cogn.
Catasetum fimbriatum Rchb.f. non Lindl.
Catasetum pusillum C. Schweinf. fide Brako, L & J.L.Zarucchi
Monachanthus fimbriatus Gardn. ex Hook.
Monachanthus roseo-albus Hook.

Distribution: Brazil, Colombia, Venezuela

Catasetum saccatum Lindl.
Catasetum baraquinianum Lem.
Catasetum christyanum Rchb.f.
Catasetum christyanum var. *chlorops* Rchb.f.
Catasetum christyanum var. *obscurum* Rchb.f.
Catasetum colossus Schltr.
Catasetum cruciatum Schltr.
Catasetum histrio Klotzsch ex Rchb.f.

Catasetum japurense Mansf.
Catasetum saccatum var. *album* Hort. ex Pabst & Dungs
Catasetum saccatum var. *chlorops* (Rchb.f.) Mansf.
Catasetum saccatum var. *christyanum* (Rchb.f.) Mansf.
Catasetum saccatum var. *eusaccatum* Mansf.
Catasetum saccatum var. *incurvum* (Klotzsch) Mansf.
Catasetum saccatum var. *pliciferum* Rchb.f.
Catasetum saccatum var. *typum* Hoehne
Catasetum saccatum var. *viride* Hoehne

Distribution: Bolivia, Brazil, Colombia, Ecuador, Guyana, Peru

Catasetum samaniegoi Dodson

Distribution: Ecuador

Catasetum sanguineum Lindl. & Paxton
Catasetum naso Hook. non Lindl.
Catasetum naso var. *pictum* T.Moore
Catasetum naso var. *viride* T.Moore
Catasetum sanguineum var. *integrale* Rchb.f.
Catasetum sanguineum var. *viride* (T.Moore) Jenny
Myanthus sanguineus Linden

Distribution: Colombia, Costa Rica, Panama, Venezuela

Catasetum schmidtianum F.E.L.Miranda & K.G.Lacerda

Distribution: Brazil

Catasetum schunkei Dodson & D.E.Benn.

Distribution: Peru

Catasetum schweinfurthii D.E.Benn. & Christenson

Distribution: Peru

Catasetum semicirculatum Miranda

Distribution: Brazil

Catasetum × sodiroi Schltr.
Catasetum chloranthum Cogn.
Catasetum platyglossum var. *sodiroi* Mansf.
Catasetum trilobatum Senghas

Distribution: Ecuador, Peru

Catasetum spinosum (Hook.) Lindl.
Myanthus spinosus Hook.

Distribution: Brazil, Colombia, Guyana, Venuzuela

Catasetum spitzii Hoehne
Catasetum spitzii var. *album* L.C.Menezes
Catasetum spitzii var. *roseum* L.C.Menezes
Catasetum spitzii var. *sanguineum* L.C.Menezes

Distribution: Bolivia, Brazil

Catasetum stevensonii Dodson

Distribution: Ecuador

Catasetum tabulare Lindl.
Catasetum caucanum Schltr.
Catasetum finetianum L.Lind & Cogn.
Catasetum pallidiflorum Schltr.
Catasetum rhamphastos Kraenzl.
Catasetum tabulare var. *brachyglossum* Rchb.f.
Catasetum tabulare var. *finetianum* Mansf.
Catasetum tabulare var. *laeve* Rchb.f.
Catasetum tabulare var. *pallidum* Mansf.
Catasetum tabulare var. *rhamphastos* Mansf.
Catasetum tabulare var. *rhinophorum* Rchb.f.
Catasetum tabulare var. *rugosum* Mansf.
Catasetum tabulare var. *serrulata* Rchb.f. & ex Regel
Catasetum tabulare var. *virens* Rchb.f.

Distribution: Colombia

Catasetum taguariense L.C.Menezes & G.J.Braem
Catasetum taguariense var. *album* L.C.Menezes

Distribution: Brazil

Catasetum × **tapiriceps** Rchb.f.
Catasetum albopurpureum L.Linden
Catasetum apertum Rolfe
Catasetum cabrutae Schnee
Catasetum cupuliforme Hort
Catasetum integerrimum var. *luteo-purpureum* (Cogn.) Mansf.
Catasetum lindenii Cogn.
Catasetum lucianii Cogn.
Catasetum macrocarpum var. *aurantiacum* Cogn.
Catasetum macrocarpum var. *chrysanthum* Hort ex W.H.G.
Catasetum macrocarpum var. *flavescens* Cogn. ex M. Gardner
Catasetum macrocarpum var. *luteo-purpureum* Cogn.
Catasetum macrocarpum var. *luteo-roseum* Linden

Catasetum maculatum var. *luteo-purpureum* Cogn.
Catasetum magnificum Hort
Catasetum mirabile Cogn.
Catasetum o'brienianum Hort. ex Gardner
Catasetum quadricolor Cogn.
Catasetum revolutum Cogn.
Catasetum semiroseum Beck
Catasetum splendens Cogn.
Catasetum splendens var. *acutipetalum* Linden
Catasetum splendens var. *albo-purpureum* Cogn.
Catasetum splendens var. *album* Linden & Cogn.
Catasetum splendens var. *aliciae* Linden & Cogn.
Catasetum splendens var. *atropurpureum* Linden & Cogn.
Catasetum splendens var. *aurantiacum* Rolfe
Catasetum splendens var. *aureo-maculatum* De Bosschere
Catasetum splendens var. *aureum* Cogn.
Catasetum splendens var. *eburneum* Rolfe
Catasetum splendens var. *flavescens* Rolfe
Catasetum splendens var. *griganii* Linden
Catasetum splendens var. *imperiale* (Linden & Cogn.) Rolfe
Catasetum splendens var. *lansbergianum* Linden
Catasetum splendens var. *lindeni* Rolfe
Catasetum splendens var. *luciani* Rolfe
Catasetum splendens var. *maculatum* Rolfe
Catasetum splendens var. *mirabile* (Cogn.) Rolfe
Catasetum splendens var. *o'brienianum* Rolfe
Catasetum splendens var. *punctatissimum* Rolfe
Catasetum splendens var. *regale* Gower
Catasetum splendens var. *revolutum* Cogn.
Catasetum splendens var. *rubrum* Linden & Cogn.
Catasetum splendens var. *semiroseum* Cogn.
Catasetum splendens var. *viride* Rolfe
Catasetum splendens var. *worthingtonianum* Rolfe

Distribution: Brazil, Colombia, Venezuela

Catasetum taquariense Bicalho, Barros & Moutinho

Distribution: Brazil

Catasetum tenebrosum Kraenzl.
Catasetum tenebrosum forma *smaragdinum* D.E.Benn., Christenson & Collantes

Distribution: Peru

Catasetum tenuiglossum Senghas

Distribution: Peru

Catasetum thompsonii Dodson

Distribution: Guyana

Catasetum tigrinum Rchb.f.
Catasetum hoehnei Mansf.

Distribution: Brazil

Catasetum transversicallosum D.E.Benn. & Christenson

Distribution: Peru

Catasetum tricorne P.Ortiz

Distribution: Colombia

Catasetum triodon Rchb.f.
Catasetum monodon Kraenzl.
Catasetum tricolor Rchb.f. (sphalm)
Catasetum triodon var. *guttulatum* Hoehne

Distribution: Brazil

Catasetum trulla Lindl.
Catasetum liechtensteinii Kraenzl.
Catasetum socco (Vell.) Hoehne
Catasetum trulla var. *liechtensteinii* Mansf.
Catasetum trulla var. *maculatissimum* Rchb.f.
Catasetum trulla var. *subimberbe* Rchb.f.
Catasetum trulla var. *trilobatum* Schltr. ex Mansf.
Catasetum trulla var. *trulla* Mansf.
Catasetum trulla var. *typum* Hoehne
Cypripedium cothurnum Vell.
Cypripedium socco Vell.
Paphiopedilum cothurnum (Vell.) Pfitzer
Paphiopedilum socco (Vell.) Pfitzer

Distribution: Brazil

Catasetum tuberculatum Dodson

Distribution: Colombia, Ecuador, Peru

Catasetum turbinatum Hoffmgg. ex Heynh.

Distribution: Brazil

Catasetum tururuiense A.T.Oliveira & da Silva

Distribution: Brazil

Part II: Catasetum

Catasetum uncatum Rolfe

Distribution: Brazil

Catasetum vibratile (Bass.) Cpm.

Distribution: Brazil

Catasetum vinaceum Hoehne
Catasetum trulla ssp. *vinaceum* Hoehne
Catasetum vinaceum var. *album* (L.C.Menezes) L.C.Menezes
Catasetum vinaceum var. *splendidum* L.C.Menezes

Distribution: Brazil

Catasetum violascens Rchb.f. & Warz.

Distribution: Peru

Catasetum viridiflavum Hook.
Catasetum serratum Lindl.

Distribution: Costa Rica, Panama

Catasetum × **wendlingeri** Foldats

Distribution: Venezuela

Catasetum yavitaense G.A.Romero & C.Gomez

Distribution: Venezuela

MILTONIA BINOMIALS IN CURRENT USE

MILTONIA BINOMES ACTUELLEMENT EN USAGE

MILTONIA BINOMIALES UTILIZADOS NORMALMENTE

Miltonia × **binotii** Cogn.

Distribution: Brazil

Miltonia × **bluntii** Rchb.f.
Miltonia peetersiana Rchb.f.

Distribution: Brazil

Miltonia × **castanea** Rolfe
Miltonia clowesii var. *castanea* Rchb.f.
Miltonia lawrenceana Cogn.

Distribution: Brazil

Miltonia clowesii Lindl.
Brassia clowesii (Lindl.) Lindl.
Odontoglossum clowesii Lindl.
Oncidium clowesii (Lindl.) Beer

Distribution: Brazil

Miltonia × **cogniauxiae** Peeters
Miltonia cogniauxiae var. *bicolor* Rolfe
Miltonia cogniauxiae var. *massaiana* Cogn.
Miltonia cogniauxiae var. *pallida* Cogn.

Distribution: Brazil

Miltonia × **festiva** (Rchb.f.) Nichols
Miltonia cyrtochiloides Barb.Rodr.

Distribution: Brazil

Miltonia flava Lindl.
Miltonia anceps Lindl.
Miltonia pinelli Hort ex Rchb.f.
Odontoglossum anceps Klotzsch, Otto & Dietrich
Oncidium anceps (Klotzsch) Rchb.f.

Distribution: Brazil

Part II: Miltonia

Miltonia flavescens (Lindl.) Lindl.
Cyrtochilum flavescens Lindl.
Cyrtochilum stellatum Lindl.
Miltonia flavescens var. *grandiflora* Regel
Miltonia flavescens var. *stellata* Regel
Miltonia flavescens var. *typica* Regel
Miltonia loddigesii Hort ex Rchb.f.
Miltonia stellata Lindl.
Oncidium flavescens (Lindl.) Rchb.f.
Oncidium stellatum (Lindl.) Beer

Distribution: Argentina, Brazil, Paraguay

Miltonia × lamarckeana Rchb.f.
Miltonia clowesii var. *lamarckeana* E.Morren
Miltonia joiceyana O'Brien

Distribution: Brazil

Miltonia × leucoglossa Hort

Distribution: Brazil

Miltonia odorata Lodd. ex W.Baxt.

Distribution: Brazil

Miltonia regnellii Rchb.f.
Miltonia cereola Lem.
Miltonia regnelli var. *travassosiana* Cogn.
Miltonia regnelli var. *veitchiana* Cogn.
Oncidium regnellii (Rchb.f.) Rchb.f.

Distribution: Brazil

Miltonia × rosina Barb.Rodr.

Distribution: Brazil

Miltonia spectabilis Lindl.
Macrochilus fryanus Knowles & Westc.
Miltonia bicolor Lodd. ex W.Baxt.
Miltonia moreliana Hort. ex Lindl.
Miltonia rosea Verschaff. ex Lem.
Miltonia spectabilis var. *lineata* Lind & Rod
Miltonia spectabilis var. *morelliana* Henfr.
Miltonia spectabilis var. *virginalis* Lem.
Miltonia warneri Nichols
Oncidium spectabile (Lindl.) Beer

Distribution: Brazil, Venezuela

MILTONIOIDES BINOMIALS IN CURRENT USE

MILTONIOIDES BINOMES ACTUELLEMENT EN USAGE

MILTONIOIDES BINOMIALES UTILIZADOS NORMALMENTE

Miltonioides carinifera (Rchb.f.) Senghas & Luckel
Odontoglossum cariniferum Rchb.f.
Oncidium cariniferum (Rchb.f.) Beer

Distribution: Costa Rica, Panama

Miltonioides karwinskii (Lindl.) Brieger & Luckel
Cyrtochilum karwinskii Lindl.
Miltonia karwinskii (Lindl.) Lindl.
Odontoglossum karwinskii (Lindl.) Rchb.f.
Odontoglossum laeve var. *karwinskii* (Lindl.) Sawyer
Oncidium karwinskii (Lindl.) Lindl.

Distribution: Mexico

Miltonioides laevis (Lindl.) Brieger & Luckel
Odontoglossum laeve Lindl.
Oncidium laeve (Lindl.) Beer

Distribution: Guatemala, Mexico

Miltonioides leucomelas (Rchb.f.) Bockemuhl & Senghas
Miltonia leucomelas (Rchb.f.) Rolfe
Miltonioides pauciflora (L.O.Williams) Hamer & Garay
Miltonioides stenoglossum (Schltr.) Brieger & Luckel
Odontoglossum laeve var. *auratum* Rchb.f.
Odontoglossum leucomelas Rchb.f.
Odontoglossum pauciflorum L.O.Williams
Odontoglossum stenoglossum (Schltr.) L.O.Williams

Distribution: Costa Rica, El Salvador, Guatemala, Honduras, Mexico, Nicaragua

Miltonioides oviedomotae (Hagsater) Senghas
Oncidium oviedomotae Hagsater

Distribution: Mexico

Miltonioides reichenheimii (Linden & Rchb.f.) Brieger & Luckel
Miltonia reichenheimii (Linden & Rchb.f.) Rolfe
Odontoglossum laeve var. *reichenheimii* (Linden & Rchb.f.) O'Brien
Odontoglossum reichenheimii Linden & Rchb.f.
Oncidium reichenheimii (Linden & Rchb.f.) Garay & Stacy

Distribution: Mexico

Part II: Miltonioides

Miltonioides schroederiana (O'Brien) Brieger & Luckel
Miltonia schroederiana O'Brien
Miltonioides confusa (Garay) Brieger & Luckel
Odontoglossum confusum Garay
Odontoglossum confusum var. *album* Hort.
Odontoglossum schroederianum Rchb.f.
Oncidium schroederianum (O'Brien) Garay & Stacy

Distribution: Costa Rica, Panama

MILTONIOPSIS BINOMIALS IN CURRENT USE

MILTONIOPSIS BINOMES ACTUELLEMENT EN USAGE

MILTONIOPSIS BINOMIALES UTILIZADOS NORMALMENTE

Miltoniopsis bismarckii Dodson & D.E.Benn.
Miltonia bismarckii (Dodson & D.E.Benn.) P.F.Hunt

Distribution: Peru

Miltoniopsis phalaenopsis (Rchb.f.) Garay & Dunst.
Miltonia phalaenopsis (Rchb.f.) Nichols
Miltonia pulchella Hort. ex Linden
Odontoglossum phalaenopsis Rchb.f.

Distribution: Colombia, Peru

Miltoniopsis roezlii (Rchb.f.) God.-Leb.
Miltonia roezlii (Rchb.f.) Nichols
Miltoniopsis santanaei Garay & Dunst. sensu Jørgensen
Odontoglossum roezlii Rchb.f.

Distribution: Colombia, Ecuador, Panama, Peru

Miltoniopsis roezlii ssp. **alba** (W.Bull ex W.G.Sm.) Luckel
Miltonia roezlii ssp. *alba* (W.Bull ex W.G.Sm.) B.S.Williams
Miltonia roezlii subvar. *alba* (W.Bull ex W.G.Sm.) Veitch
Odontoglossum roezlii var. *album* W.Bull ex W.G.Sm.

Distribution: Venezuela

Miltoniopsis santanaei Garay & Dunst.

Distribution: Colombia, Ecuador, Peru, Venezuela

Miltoniopsis vexillaria (Rchb.f.) God.-Leb.
Miltonia vexillaria (Rchb.f.) Nichols
Miltonia vexillaria var. *alba* A.Echavarria & O.Robledo
Miltonia vexillaria var. *carolina* A.Echavarria & O.Robledo
Miltonia vexillaria var. *hilliana* A.Echavarria & O.Robledo
Miltonia vexillaria var. *kienastiana* A.Echavarria & O.Robledo
Miltonia vexillaria var. *leopoldii* Veitch
Miltonia vexillaria var. *leucoglossa* A.Echavarria & O.Robledo
Miltonia vexillaria var. *stupenda* Veitch
Miltonia vexillaria var. *superba* A.Echavarria & O.Robledo
Miltonia vexillaria var. *rosea* A.Echavarria & O.Robledo
Miltonia vexillaria var. *rubella* Veitch

Odontoglossum vexillarium Rchb.f.

Distribution: Colombia, Costa Rica, Ecuador, Peru

Miltoniopsis warscewiczii (Rchb.f.) Garay & Dunst.
Miltonia endresii Nichols
Miltonia superba Schltr.
Odontoglossum warscewiczianum Rchb.f. ex Hemsl.
Odontoglossum warscewiczii Rchb.f.

Distribution: Costa Rica, Panama, Venezuela

RENANTHERA BINOMIALS IN CURRENT USE

RENANTHERA BINOMES ACTUELLEMENT EN USAGE

RENANTHERA BINOMIALES UTILIZADOS NORMALMENTE

Renanthera amabilis Boxall ex Náves (nomen)

Distribution: Philippines

Renanthera annamensis Rolfe
Renanthera hennisiana Schltr.
Renanthera pulchella Rolfe

Distribution: Myanmar, Viet Nam

Renanthera bella J.J.Wood

Distribution: Malaysia

Renanthera citrina Aver.

Distribution: Viet Nam

Renanthera coccinea Lour.
Epidendrum renanthera Raeusch.
Gongora philippica Llanos sensu Schltr.
Renanthera coccinea var. *holttumii* Mahyar

Distribution: China, Lao People's Democratic Republic (The), Myanmar, Thailand, Viet Nam

Renanthera edefeldtii F.Muell. & Kraenzl. ex Kraenzl.

Distribution: Indonesia, Papua New Guinea, Solomon Islands

Renanthera elongata (Blume) Lindl.
Aerides elongata (Blume) Lindl.
Porphyrodesme elongata (Blume) Garay
Renanthera matutina auct. non (Blume) Lindl.
Renanthera micrantha Blume
Saccolabium reflexum Lindl.

Distribution: Indonesia, Malaysia, Philippines (The), Singapore, Thailand

Renanthera imschootiana Rolfe
Renanthera matutina Lindl. sensu Rolfe

Part II: Renanthera

Renanthera papilio King & Pantl.

Distribution: China, India, Myanmar, Viet Nam

Renanthera isosepala Holttum
Renanthera matutina auct. non (Blume) Lindl.

Distribution: Indonesia, Malaysia, Thailand

Renanthera matutina Lindl.
Aerides matutina Blume (non Wild.)
Nephranthera matutina Hassk.
Renanthera angustifolia Hook.f. sensu Ridl.

Distribution: Indonesia, Philippines (The), Thailand

Renanthera moluccana Blume
Angraecum rubrum Rumph.

Distribution: Indonesia, Papua New Guinea

Renanthera monachica Ames

Distribution: Philippines (The)

Renanthera philippinensis (Ames & Quisumb.) L. O.Williams
Renanthera storiei var. *philippinensis* Ames & Quisumb.

Distribution: Papua New Guinea, Philippines (The)

Renanthera storiei Rchb.f.
Renanthera storiei forma *citrina* Valmayor & D.Tiu
Vanda storiei Storie ex Rchb.f.

Distribution: Philippines (The)

Renanthera striata Rolfe

Distribution: Colombia, Ecuador, Panama, Venezuela

Renanthera sulingi Lindl.
Aerides sulingi Blume
Armodorum sulingi (Blume) Schltr.
Vanda sulingi Blume

Distribution: Indonesia

RENANTHERELLA BINOMIALS IN CURRENT USE

RENANTHERELLA BINOMES ACTUELLEMENT EN USAGE

RENANTHERELLA BINOMIALES UTILIZADOS NORMALMENTE

Renantherella histrionica (Rchb.f.) Ridl.
Renanthera auyongi Christenson
Renanthera histrionica Rchb.f.

Distribution: Malaysia, Thailand

RHYNCHOSTYLIS BINOMIALS IN CURRENT USE

RHYNCHOSTYLIS BINOMES ACTUELLEMENT EN USAGE

RHYNCHOSTYLIS BINOMIALES UTILIZADOS NORMALMENTE

Rhynchostylis coelestis Rchb.f.
Saccolabium coeleste Rchb.f.
Vanda pseudo-caerulescens Guill.

Distribution: Cambodia, Lao People's Democratic Republic (The), Thailand, Viet Nam

Rhynchostylis gigantea (Lindl.) Ridl.
Anota densiflora (Lindl.) Schltr.
Anota gigantea (Lindl.) Fukuy.
Anota hainanensis (Rolfe) Schltr.
Anota harrisoniana (Hook.) Schltr.
Rhynchostylis densiflora (Lindl.) L.O.Williams
Rhynchostylis gigantea subvar. *petotiana* (Rchb.f.) Guill.
Rhynchostylis gigantea var. *harrisoniana* (Hook.) Holttum
Saccolabium albolineatum Teijsm. & Binn.
Saccolabium giganteum Lindl.
Saccolabium harrisonianum Hook.
Saccolabium violaceum auct. non Rchb.f.
Vanda densiflora Lindl.
Vanda densiflora var. *petotiana* Rchb.f.
Vanda hainanensis Rolfe

Distribution: Cambodia, China, Indonesia, Lao People's Democratic Republic (The), Malaysia, Myanmar, Singapore, Thailand, Viet Nam

Rhynchostylis gigantea ssp. **violacea** (Lindl.) Christenson
Rhynchostylis violacea Rchb.f.
Saccolabium violaceum Rchb.f.
Vanda violacea Lindl.

Distribution: Philippines (The)

Rhynchostylis retusa (L.) Blume
Aerides guttata (Lindl.) Roxb.
Aerides praemorsa Willd.
Aerides retusum (L.) Sw.
Aerides spicata D.Don
Aerides undulatum Sm.
Epidendrum hippium Buch.
Epidendrum indicum Poir.
Epidendrum retusum L.
Gastrochilus blumei (Lindl.) Kuntze
Gastrochilus gurwalicus (Lindl.) Kuntze
Gastrochilus praemorsus (Willd.) Dur. & Jacks.
Gastrochilus retusus (L.) Kuntze
Gastrochilus rheedii (Wight) Kuntze

Gastrochilus spicatus (D.Don) Kuntze
Limodorum retusa (L.) Sw.
Orchis lanigera Blanco
Rhynchostylis gurwalica (Lindl.) Rchb.f.
Rhynchostylis guttata (Lindl.) Rchb.f.
Rhynchostylis praemorsa (Willd.) Blume
Rhynchostylis retusa ssp. *macrostachya* Rchb.f.
Rhynchostylis violacea auct. non Rchb.f.
Rhynchostylis violacea var. *berkeleyi* (Rchb.f.) Stein
Saccolabium berkeleyi Rchb.f.
Saccolabium blumei Lindl.
Saccolabium blumei ssp. *major* Williams
Saccolabium blumei var. *russelianum* Williams
Saccolabium furcatum Hort
Saccolabium gurwalicum Lindl.
Saccolabium guttatum (Lindl.) Lindl.
Saccolabium heathii Hort
Saccolabium holfordianum Hort
Saccolabium littorale Rchb.f.
Saccolabium macrostachyum Lindl.
Saccolabium praemorsum (Willd.) Lindl.
Saccolabium retusum (Sw.) Voigt
Saccolabium rheedii Wight
Saccolabium spicatum (D.Don) Lindl.
Saccolabium turneri Williams
Sarcanthus guttatus Lindl.

Distribution: Bhutan, Cambodia, China, India, Indonesia, Lao People's Democratic Republic (The), Malaysia, Myanmar, Nepal, Philippines (The), Sri Lanka, Thailand, Viet Nam

ROSSIOGLOSSUM BINOMIALS IN CURRENT USE

ROSSIOGLOSSUM BINOMES ACTUELLEMENT EN USAGE

ROSSIOGLOSSUM BINOMIALES UTILIZADOS NORMALMENTE

Rossioglossum grande (Lindl.) Garay & G.C.Kenn.
Odontoglossum grande Lindl.
Odontoglossum grande var. *excelsior* Kerch.
Odontoglossum grande var. *hibernum* Hort
Odontoglossum grande var. *magnificum* B.S.Williams
Odontoglossum grande var. *splendens* Rchb.f.
Odontoglossum grande var. *superbum* B.S.Williams

Distribution: Guatemala, Mexico

Rossioglossum grande var. **aureum** (Hort. ex Stein) Garay & G.C.Kenn.
Odontoglossum grande var. *aureum* Hort. ex Stein
Odontoglossum grande var. *citrinum* Hort.
Odontoglossum grande var. *pittianum* Hort
Odontoglossum grande var. *sanderae* O'Brien

Distribution: Guatemala, Mexico

Rossioglossum insleayi (Barker ex Lindl.) Garay & G.C.Kenn.
Odontoglossum insleayi (Barker ex Lindl.) Lindl.
Odontoglossum lawrenceanum Hort. ex Gard.
Oncidium insleayi Barker ex Lindl.

Distribution: Mexico

Rossioglossum powellii (Schltr.) Garay & G.C.Kenn.
Odontoglossum powellii Schltr.
Odontoglossum schlieperianum var. *pretiosum* Rchb.f.

Distribution: Costa Rica, Panama

Rossioglossum schlieperianum (Rchb.f.) Garay & G.C.Kenn.
Odontoglossum insleayi var. *macranthum* Lindl.
Odontoglossum lawrenceanum Hort.
Odontoglossum schlieperianum Rchb.f.
Odontoglossum warscewiczii sensu Bridges

Distribution: Costa Rica, Panama

Rossioglossum schlieperianum var. **flavidum** (Rchb.f.) Garay & G.C.Kenn.
Odontoglossum grande var. *flavidum* Klotzsch ex Rchb.f.
Odontoglossum grande var. *pallidum* Klotzsch ex Rchb.f.
Odontoglossum schlieperianum ssp. *actuinum* O'Brien
Odontoglossum schlieperianum var. *citrinum* O'Brien

Odontoglossum schlieperianum var. *flavidum* Rchb.f.
Odontoglossum schlieperianum var. *xauthinum* Baliff.

Distribution: Costa Rica, Panama

Rossioglossum splendens (Rchb.f.) Garay & G.C.Kenn.
Odontoglossum insleayi var. *splendens* Rchb.f.
Odontoglossum splendens (Rchb.f.) G.C.Kenn.

Distribution: Mexico

Rossioglossum splendens ssp. **imschootianum** (Rolfe) Garay & G.C.Kenn.
Odontoglossum insleayi var. *aureum* B.S.Williams
Odontoglossum insleayi var. *imschootianum* Rolfe

Distribution: Mexico

Rossioglossum splendens var. **leopardinum** (Regel) Garay & G.C.Kenn.
Odontoglossum insleayi var. *leopardinum* Regel
Onciduim leopardinum Kerch. (non Lindl.)

Distribution: Mexico

Rossioglossum splendens ssp. **pantherinum** (Rchb.f.) Garay & G.C.Kenn.
Odontoglossum insleayi var. *pantherinum* Rchb.f.

Distribution: Mexico

Rossioglossum williamsianum (Rchb.f.) Garay & G.C.Kenn.
Odontoglossum grande var. *williamsianum* (Rchb.f.) Veitch
Odontoglossum williamsianum Rchb.f.

Distribution: Costa Rica, Guatemala, Honduras

VANDA BINOMIALS IN CURRENT USE

VANDA BINOMES ACTUELLEMENT EN USAGE

VANDA BINOMIALES UTILIZADOS NORMALMENTE

Vanda alpina (Lindl.) Lindl.
Luisia alpina Lindl.
Trudelia alpina (Lindl.) Garay

Distribution: Bhutan, China, India, Nepal

Vanda arbuthnotiana Kraenzl.

Distribution: India

Vanda arcuata J.J.Sm.

Distribution: Indonesia

Vanda bensonii Bateman

Distribution: Myanmar, Thailand

Vanda bicolor Griff.

Distribution: Bhutan

Vanda bidupensis Aver. & Christenson

Distribution: Viet Nam

Vanda × boumaniae J. J. Smith

Distribution: Indonesia

Vanda brunnea Rchb.f.
Vanda denisoniana var. *hebraica* Rchb.f.
Vanda henryi Schltr.

Distribution: China, Myanmar, Thailand

Vanda celebica Rolfe

Distribution: Indonesia

Vanda × charlesworthii Rolfe

Distribution: Myanmar

Vanda chlorosantha (Garay) Christenson
Trudelia chlorosantha Garay

Distribution: Bhutan

Vanda coerulea Griff. ex Lindl.
Vanda coerulea var. *concolor* Cogn.
Vanda coerulea ssp. *hennisiane* Schltr.
Vanda coerulea ssp. *sanderae* Rchb.f.
Vanda coerulescens Lindl.

Distribution: China, India, Myanmar, Thailand

Vanda coerulescens Griff.

Distribution: China, India, Myanmar, Thailand

Vanda coerulescens ssp. **boxallii** Rchb.f.

Distribution: ?Myanmar, Nepal

Vanda concolor Blume
Angraecum furvum Rumph.
Cymbidium furvum Willd.
Epidendrum furvum L.
Vanda esquirolei Schltr.
Vanda furva Lindl.
Vanda fuscoviridis Lindl.
Vanda guangxiensis Fowlie
Vanda kwangtungensis S.J.Cheng & C.Z.Tang
Vanda lindenii auct. non Rchb.f.
Vanda roxburgii var. *unicolor* Hook
Vanda stella Hort

Distribution: China, Lao People's Democratic Republic (The), Viet Nam

Vanda × confusa Rolfe

Distribution: Myanmar

Vanda crassiloba Teijsm. & Binn.

Distribution: Indonesia

Vanda cristata Lindl.
Trudelia cristata (Lindl.) Senghas
Vanda striata Rchb.f.

Distribution: Bangladesh, China, India, Myanmar, Nepal

Vanda cruenta Lodd. Cat. ex Sweet

Distribution: China

Vanda dearei Rchb.f.

Distribution: Indonesia, Malaysia

Vanda denisoniana Benson & Rchb.f.
Vanda denisoniana var. *tessellata* Guill.
Vanda micholitzii Rolfe ex Ridl.
Vanda suavis auct. non Lindl.
Vanda tricolor auct. non Lindl.

Distribution: Lao People's Democratic Republic (The), Myanmar, Thailand, Viet Nam

Vanda devoogtei J.J.Sm.

Distribution: Indonesia

Vanda flavobrunnea Rchb.f.

Distribution: Malaysia

Vanda foetida J.J.Sm.

Distribution: Indonesia

Vanda griffithii Lindl.
Trudelia griffithii (Lindl.) Garay

Distribution: Bhutan

Vanda hastifera Rchb.f.
Renanthera trichoglottis Ridl.
Vanda gibbsiae Rolfe
Vanda hastifera var. *gibbsiae* (Rolfe) P.J.Cribb

Distribution: Indonesia, Malaysia

Part II: Vanda

Vanda helvola Blume

Distribution: Indonesia, Malaysia, Papua New Guinea

Vanda hindsii Lindl.
Vanda suavis F.Muell.
Vanda truncata J.J.Sm.
Vanda whiteana D.A.Herb. & S.T.Blake

Distribution: Australia, Indonesia, Papua New Guinea, Solomon Islands

Vanda insignis Blume
Vanda insignis var. *schroederiana* Rchb.f.

Distribution: Indonesia, Malaysia, Singapore

Vanda jainii A.S.Chauhan

Distribution: India

Vanda javierae D.Tiu ex Fessel & Luckel

Distribution: Philippines (The)

Vanda lamellata Lindl.
Vanda amiensis Masam. & Segawa
Vanda boxallii Rchb.f.
Vanda boxallii ssp. *cobbiana* Rchb.f.
Vanda clitellaria Rchb.f.
Vanda cumingii Lodd.
Vanda lamellata forma *flava* Valmayor & D.Tiu
Vanda lamellata var. *boxallii* Rchb.f.
Vanda lamellata var. *calayana* Valmayor & D.Tiu
Vanda lamellata var. *cobbiana* (Rchb.f.) Ames
Vanda lamellata var. *remediosae* Ames & Quisumb.
Vanda nasughuana Parsons
Vanda superba Linden & Rodr.
Vanda unicolor sensu IK
Vanda vidalii Boxall ex Naves
Vanda yamiensis Masam. & Segawa

Distribution: China, Indonesia, Japan, Malaysia, Philippines (The)

Vanda leucostele Schltr.

Distribution: Indonesia

Vanda lilacina Teijsm. & Binn.
Aerides wightiana auct. non Lindl.

Pteroceras caligare auct. non (Ridl.) Holttum
Sarcochilus caligaris auct. non Ridl.
Vanda laotica Guill.
Vanda parviflora auct. non Lindl.
Vanda parviflora var. *albiflora* Hook.f.

Distribution: Cambodia, China, Lao People's Democratic Republic (The), Myanmar, Thailand, Viet Nam

Vanda limbata Blume

Distribution: Indonesia, Philippines (The)

Vanda lindeni Rchb.f.
Vanda saxatilis J.J.Sm.

Distribution: Indonesia, Philippines (The), Viet Nam

Vanda liouvillei Finet
Vanda brunnea auct. non Rchb.f.

Distribution: Cambodia, Indonesia, Lao People's Democratic Republic (The), Malaysia, Myanmar, Thailand, Viet Nam

Vanda lombokensis J.J.Sm.

Distribution: Indonesia

Vanda luzonica Loher ex Rolfe
Vanda tricolor Ames non Lindl.

Distribution: Indonesia, Philippines (The)

Vanda merrillii Ames & Quisumb.
Vanda merrillii var. *immaculata* Ames & Quisumb.
Vanda merrillii var. *rotorii* Ames & Quisumb.

Distribution: Philippines (The)

Vanda moorei Rolfe

Distribution: Indonesia, Malaysia, Myanmar

Vanda pauciflora Breda

Distribution: Indonesia

Part II: Vanda

Vanda pumila Hook.f.
Trudelia pumila (J.D.Hook.) Senghas

Distribution: China, India, Indonesia, Lao People's Democratic Republic (The), Nepal, Thailand, Viet Nam

Vanda punctata Ridl.

Distribution: Indonesia, Malaysia

Vanda roeblingiana Rolfe

Distribution: Philippines (The)

Vanda sanderiana Rchb.f.
Esmerelda sanderiana Rchb.f.
Esmerelda sanderiana var. *albata* Will.
Esmerelda sanderiana var. *labello-viridi* (Linden & Rodr.) Will.
Euanthe sanderiana (Rchb.f.) Schltr.
Vanda sanderiana var. *albata* Rchb.f.
Vanda sanderiana var. *froebelliana* Cogn.
Vanda sanderiana var. *labello-viridi* Linden & Rodr.

Distribution: Philippines (The)

Vanda scandens Holttum

Distribution: Indonesia, Malaysia

Vanda spathulata (L.) Spreng.
Aerides maculatum Buch.-Ham
Aerides tessellatum Wight in Wall.
Cymbidium allagnata Buch.-Ham. ex Wall.
Cymbidium spatulatum Moon
Epidendrum spathulatum L.
Epidendrum tesseloides Steud.
Limodorum spathulatum (L.) Willd.
Taprobanea spathulata (L.) Christenson

Distribution: India, Sri Lanka

Vanda stangeana Rchb.f.
Vanda petersiana Schltr.

Distribution: India, Nepal

Vanda subconcolor Tang & F.T.Wang
Vanda subconcolor var. *disticha* Tang & F.T.Wang

Distribution: China

Vanda sumatrana Schltr.

Distribution: Indonesia

Vanda taiwaniana S.S.Ying

Distribution: China

Vanda tessellata (Roxb.) Hook. ex G.Don
Cymbidium tesselatum (Roxb.) Sw.
Cymbidium tesseloides Roxb.
Epidendrum tesselatum Roxb.
Vanda roxburghii R.Br.
Vanda tessellata var. *lutescens* M.E. Dalpethado
Vanda tessellata var. *rufescens* M.E. Dalpethado
Vanda tesselloides (Roxb.) Rchb.f.
Vanda unicolor Steud. sensu Christenson

Distribution: India, Myanmar, Nepal, Sri Lanka

Vanda testacea (Lindl.) Rchb.f.
Aerides orthocentra Hand.-Mazz.
Aerides testacea Lindl.
Aerides wightiana Lindl.
Vanda parviflora Lindl.
Vanda spathulata auct. non (L.) Spreng.
Vanda vitellina Kraenzl.

Distribution: India, Myanmar, Nepal, Sri Lanka

Vanda thwaitesii Hook.f.
Aerides tesselatum Thw. non Wight

Distribution: Sri Lanka

Vanda tricolor Lindl.
Vanda hindsii Benth. non Lindl.
Vanda suaveolens Blume
Vanda tricolor forma *patersonii* (Hort.) Hiroe
Vanda tricolor var. *patersonii* Hort.

Distribution: Australia, Indonesia

Part II: Vanda

Vanda tricolor ssp. **suavis** (Lindl.) Veitch
Vanda suavis Lindl.

Distribution: Indonesia, Lao People's Democratic Republic (The)

Vanda vipanii Rchb.f.

Distribution: Myanmar

Vanda wightii Rchb.f.

Distribution: India

VANDOPSIS BINOMIALS IN CURRENT USE

VANDOPSIS BINOMES ACTUELLEMENT EN USAGE

VANDOPSIS BINOMIALES UTILIZADOS NORMALMENTE

Vandopsis gigantea (Lindl.) Pfitzer
Fieldia gigantea (Lindl.) Rchb.f.
Stauropsis chinensis Rolfe
Stauropsis gigantea (Lindl.) Benth. & Hook.f.
Vanda gigantea Lindl.
Vanda lindleyana Griff. ex Lindl. & Paxton
Vandopsis chinensis (Rolfe) Schltr.

Distribution: China, Lao People's Democratic Republic (The), Malaysia, Myanmar, Thailand, Viet Nam

Vandopsis lissochiloides (Gaudich.) Pfitzer
Fieldia lissochiloides Gaudich.
Grammatophyllum pantherinum Zipp. ex Bl.
Stauropsis batemanii (Lindl.) Nicols.
Stauropsis gigantea auct. non (Lindl.) Benth. & Hook.f.
Stauropsis lissochiloides (Gaudich.) Benth. ex Pfitzer
Vanda batemanii Lindl.
Vanda lissochiloides (Gaudich.) Lindl.

Distribution: Indonesia, Lao People's Democratic Republic (The), Papua New Guinea, Philippines (The), Thailand

Vandopsis muelleri (Kraenzl.) Schltr.
Arachnis beccarii (Rchb.f.) J.J.Sm.
Arachnis muelleri (Kraenzl.) J.J.Sm.
Sarcanthopsis muelleri (Kraenzl.) Garay
Vanda muelleri Kraenzl.
Vandopsis beccarii J.J.Sm.

Distribution: Papua New Guinea

Vandopsis shanica (Phillimore & Sm.) Garay
Stauropsis shanica Phillimore & Sm.

Distribution: Myanmar

Vandopsis undulata (Lindl.) J.J.Sm.
Fieldia undulata Rchb.f.
Stauropsis polyantha W.W.Sm.
Stauropsis undulata (Lindl.) Benth. ex Hook.f.
Vanda undulata Lindl.

Distribution: Bhutan, China, India, Nepal

PART III: COUNTRY CHECKLIST
For the genera:

Aerangis, Angraecum, Ascocentrum, Bletilla, Brassavola, Calanthe, Catasetum, Miltonia, Miltonioides, Miltoniopsis, Renanthera, Renantherella, Rhynchostylis, Rossioglossum, Vanda and *Vandopsis*

TROISIEME PARTIE: LISTE PAR PAYS
Pour les genre:

Aerangis, Angraecum, Ascocentrum, Bletilla, Brassavola, Calanthe, Catasetum, Miltonia, Miltonioides, Miltoniopsis, Renanthera, Renantherella, Rhynchostylis, Rossioglossum, Vanda et *Vandopsis*

PARTE III: LISTA POR PAISES
Para el genero:

Aerangis, Angraecum, Ascocentrum, Bletilla, Brassavola, Calanthe, Catasetum, Miltonia, Miltonioides, Miltoniopsis, Renanthera, Renantherella, Rhynchostylis, Rossioglossum, Vanda y *Vandopsis*

Part III: Country Checklist / Liste par Pays / Lista por Paises

Part III: Country checklist for the genera:
Aerangis, Angraecum, Ascocentrum, Bletilla, Brassavola, Calanthe, Catasetum, Miltonia, Miltonioides, Miltoniopsis, Renanthera, Renantherella, Rhynchostylis, Rossioglossum, Vanda and *Vandopsis*

Troisième partie: Liste par pays Pour les genre:
Aerangis, Angraecum, Ascocentrum, Bletilla, Brassavola, Calanthe, Catasetum, Miltonia, Miltonioides, Miltoniopsis, Renanthera, Renantherella, Rhynchostylis, Rossioglossum, Vanda et *Vandopsis*

Parte III: Lista por paises Para el genero:
Aerangis, Angraecum, Ascocentrum, Bletilla, Brassavola, Calanthe, Catasetum, Miltonia, Miltonioides, Miltoniopsis, Renanthera, Renantherella, Rhynchostylis, Rossioglossum, Vanda y *Vandopsis*

ANGOLA / ANGOLA (L') /ANGOLA

Aerangis brachycarpa (A.Rich.) Dur. & Schinz
Aerangis calantha (Schltr.) Schltr.
Aerangis verdickii (De Wild.) Schltr.
Angraecum distichum Lindl.
Angraecum eichlerianum Kraenzl.
Calanthe sylvatica (Thouars) Lindl.

ARGENTINA / ARGENTINE (L') / ARGENTINA

Brassavola perrinii Lindl.
Miltonia flavescens (Lindl.) Lindl.

AUSTRALIA / AUSTRALIE (L') / AUSTRALIA

Calanthe triplicata (Willem.) Ames
Vanda hindsii Lindl.
Vanda tricolor Lindl.

BANGLADESH / BANGLADESH (L') / BANGLADESH

Ascocentrum ampullaceum (Roxb.) Schltr.
Vanda cristata Lindl.

BELIZE / BELIZE (LE) / BELICE

Brassavola acaulis Lindl.
Brassavola cucullata (L.) R.Br.
Brassavola nodosa (L.) Lindl.
Catasetum integerrimum Hook.

BENIN / BÉNIN (LE) / BENIN

Angraecum distichum Lindl.

BHUTAN / BHOUTAN (LE) / BHUTÁN

Ascocentrum ampullaceum (Roxb.) Schltr.
Ascocentrum himalaicum (Deb., Sengupta & Malick) Christenson
Calanthe alismaefolia Lindl.
Calanthe alpina Hook.f.
Calanthe brevicornu Lindl.
Calanthe chloroleuca Lindl.
Calanthe griffithii Lindl.
Calanthe mannii Hook.f.
Calanthe odora Griff.
Calanthe pachystalix Rchb.f.
Calanthe plantaginea Lindl.
Calanthe sylvatica (Thouars) Lindl.
Calanthe tricarinata Lindl.
Calanthe triplicata (Willem.) Ames
Rhynchostylis retusa (L.) Blume
Vanda alpina (Lindl.) Lindl.
Vanda bicolor Griff.
Vanda chlorosantha (Garay) Christenson
Vanda griffithii Lindl.
Vandopsis undulata (Lindl.) J.J.Sm.

BOLIVIA / BOLIVIE (LA) / BOLIVIA

Brassavola cebolleta Rchb.f.
Brassavola chacoensis Kraenzl.
Brassavola martiana Lindl.
Brassavola perrinii Lindl.
Catasetum buchtienii Kraenzl.
Catasetum cochabambanum Dodson & Vasquez
Catasetum duplicisculatum Senghas
Catasetum fimbriatum (Morren) Lindl. & Paxton
Catasetum fuchsii Dodson & Vasquez
Catasetum justinianum Vasquez & Dodson
Catasetum kempfii Dodson & Vasquez
Catasetum ricii Vasquez & Dodson
Catasetum rooseveltianum Hoehne
Catasetum saccatum Lindl.
Catasetum spitzii Hoehne

BRAZIL / BRÉSIL (LE) / BRASIL (EL)

Brassavola angustata Lindl.
Brassavola cebolleta Rchb.f.
Brassavola filifolia Lind.
Brassavola flagellaris Barb.Rodr.
Brassavola gardneri Cogn.
Brassavola martiana Lindl.
Brassavola nodosa (L.) Lindl.
Brassavola perrinii Lindl.

BRAZIL (continued)

Brassavola retusa Lindl.
Brassavola tuberculata Hook.
?Brassavola venosa Lindl.
Catasetum aculeatum F.E.L.Miranda & K.G.Lacerda
Catasetum albovirens Barb.Rodr.
Catasetum aripuanense Bicalho
Catasetum atratum Lindl.
Catasetum barbatum (Lindl.) Lindl.
Catasetum bergoldianum Foldats
Catasetum blackii Pabst
Catasetum boyi Mansf.
Catasetum brachybulbon Schltr.
Catasetum callosum Lindl.
Catasetum caputinum da Silva
Catasetum carolinianum F.E.Miranda & K.G.Lacerda
Catasetum cernuum (Lindl.) Rchb.f.
Catasetum collare Cogn.
Catasetum complanatum F.E.Miranda & K.G.Lacerda
Catasetum confusum G.A.Romero
Catasetum costatum Rchb.f.
Catasetum cristatum Lindl.
Catasetum cucullatum A.T.Oliveira & da Silva
Catasetum deltoideum (Lindl.) Mutel
Catasetum denticulatum Miranda
Catasetum discolor (Lindl.) Lindl.
Catasetum × dunstervillei G.A.Romero & Carnevali
Catasetum faustii Hort. ex da Silva
Catasetum ferox Kraenzl.
Catasetum fimbriatum (Morren) Lindl. & Paxton
Catasetum franchinianum K.G.Lacerda
Catasetum galeatum K.G.Lacerda
Catasetum galeritum Rchb.f.
Catasetum garnettianum Rolfe
Catasetum georgii Mansf.
Catasetum gladiatorium K.G.Lacerda
Catasetum globiflorum Hook.
Catasetum gnomus Lind. & Rchb.f.
Catasetum hookeri Lindl.
Catasetum huebneri Schltr.
Catasetum imperiale L.Lind. & Cogn.
Catasetum × intermedium Menezes & Braem
Catasetum × issanensis Pabst
Catasetum juruense Hoehne
Catasetum kleberianum Braga
Catasetum kraenzlinianum Mansf.
Catasetum lanceanum Miranda
Catasetum lanciferum Lindl.
Catasetum lemosii Rolfe
Catasetum linguiferum Schltr.

BRAZIL (continued)

Catasetum longifolium Lindl.
Catasetum longipes F.E.Miranda & K.G.Lacerda
Catasetum luridum (Link) Lindl.
Catasetum macrocarpum Rich. ex Kunth
Catasetum maranhense K.G.Lacerda & De Silva
Catasetum matogrossense Bicalho
Catasetum mattosianum Bicalho
Catasetum meeae Pabst
Catasetum micranthum Barb.Rodr.
Catasetum mocuranum Schltr.
Catasetum mojuense A.T.Oliveira & da Silva
Catasetum multifidum Miranda
Catasetum ornithoides Pabst
Catasetum osculatum K.G.Lacerda & V.P.Castro
Catasetum parguazense G.A.Romero & Carnevali
Catasetum pileatum Rchb.f.
Catasetum planiceps Lindl.
Catasetum × **pohlianum** G.Castro & Campacci
Catasetum polydactylon Schltr.
Catasetum pulchrum N.E.Br.
Catasetum punctatum Rolfe
Catasetum purum Nees & Sinning
Catasetum quadridens Rolfe
Catasetum randii Rolfe
Catasetum regnellii Barb.Rodr.
Catasetum reichenbachianum Mansf.
Catasetum richteri Bicalho
Catasetum rivularium Barb.Rodr.
Catasetum rolfeanum Mansf.
Catasetum rondonense Pabst
Catasetum rooseveltianum Hoehne
Catasetum × **roseo-album** (Hook.) Lindl.
Catasetum saccatum Lindl.
Catasetum schmidtianum F.E.L.Miranda & K.G.Lacerda
Catasetum semicirculatum Miranda
Catasetum spinosum (Hook.) Lindl.
Catasetum spitzii Hoehne
Catasetum taguariense L.C.Menezes & G.J.Braem
Catasetum × **tapiriceps** Rchb.f.
Catasetum taquariense Bicalho, Barros & Moutinho
Catasetum tigrinum Rchb.f.
Catasetum triodon Rchb.f.
Catasetum trulla Lindl.
Catasetum turbinatum Hoffmgg. ex Heynh.
Catasetum tururuiense A.T.Oliveira & da Silva
Catasetum uncatum Rolfe
Catasetum vibratile (Bass.) Cpm.
Catasetum vinaceum Hoehne
Miltonia × **binotii** Cogn.

BRAZIL (continued)

Miltonia × **bluntii** Rchb.f.
Miltonia × **castanea** Rolfe
Miltonia clowesii Lindl.
Miltonia × **cogniauxiae** Peeters
Miltonia × **festiva** (Rchb.f.) Nichols
Miltonia flava Lindl.
Miltonia flavescens (Lindl.) Lindl.
Miltonia × **lamarckeana** Rchb.f.
Miltonia × **leucoglossa** Hort
Miltonia odorata Lodd. ex W.Baxt.
Miltonia regnellii Rchb.f.
Miltonia × **rosina** Barb.Rodr.
Miltonia spectabilis Lindl.

BURUNDI / BURUNDI (LE) / BURUNDI

Aerangis kotschyana (Rchb.f.) Schltr.
Aerangis ugandensis Summerh.
Angraecum evrardianum Geerinck
Angraecum sacciferum Lindl.
Calanthe sylvatica (Thouars) Lindl.

CAMBODIA / CAMBODGE (LE) / CAMBOYA

Ascocentrum garayi Christenson
Calanthe angusta Lindl.
Calanthe lyroglossa Rchb.f.
Calanthe odora Griff.
Calanthe poilanei Gagnep.
Calanthe succedanea Gagnep.
Calanthe triplicata (Willem.) Ames
Rhynchostylis coelestis Rchb.f.
Rhynchostylis gigantea (Lindl.) Ridl.
Rhynchostylis retusa (L.) Blume
Vanda lilacina Teijsm. & Binn.
Vanda liouvillei Finet

CAMEROON / CAMEROUN (LE) / CAMERÚN (EL)

Aerangis arachnopus (Rchb.f.) Schltr.
Aerangis biloba (Lindl.) Schltr.
Aerangis calantha (Schltr.) Schltr.
Aerangis collum-cygni Summerh.
Aerangis gracillima (Kraenzl.) J.C.Arends & J.Stewart
Aerangis gravenreuthii (Kraenzl.) Schltr.
Aerangis luteo-alba (Kraenzl.) Schltr.
Aerangis luteo-alba var. **rhodosticta** (Kraenzl.) J.Stewart
Aerangis megaphylla Summerh.
Aerangis stelligera Summerh.

CAMEROON (continued)

Angraecum affine Schltr.
Angraecum angustipetalum Rendle
Angraecum aporoides Summerh.
Angraecum bancoense Van der Burg
Angraecum birrimense Rolfe
Angraecum curvipes Schltr.
Angraecum distichum Lindl.
Angraecum eichlerianum Kraenzl.
Angraecum eichlerianum var. **curvicalcaratum** Szl. & Olzs.
Angraecum firthii Summerh.
Angraecum infundibulare Lindl.
Angraecum moandense De Wild.
Angraecum podochiloides Schltr.
Angraecum pungens Schltr.
Angraecum reygaertii De Wild.
Angraecum sacciferum Lindl.
Angraecum sanfordii P.J.Cribb & B.J.Pollard
Angraecum subulatum Lindl.
Calanthe sylvatica (Thouars) Lindl.

CENTRAL AFRICAN REPUBLIC (THE) / RÉPUBLIQUE CENTRAFRICAINE (LA) / REPÚBLICA CENTROAFRICANA (LA)

Aerangis bouarensis Chiron
Aerangis calantha (Schltr.) Schltr.
Aerangis collum-cygni Summerh.
Aerangis kotschyana (Rchb.f.) Schltr.
Aerangis luteo-alba (Kraenzl.) Schltr.
Aerangis luteo-alba var. **rhodosticta** (Kraenzl.) J.Stewart
Aerangis megaphylla Summerh.
Aerangis stelligera Summerh.
Angraecum distichum Lindl.

CHINA / CHINE (LA) / CHINA

Ascocentrum ampullaceum (Roxb.) Schltr.
Ascocentrum himalaicum (Deb., Sengupta & Malick) Christenson
Ascocentrum pumilum (Hayata) Schltr.
Bletilla formosana (Hayata) Schltr.
Bletilla ochracea Schltr.
Bletilla sinensis (Rolfe) Schltr.
Bletilla striata (Thunb. ex A.Murray) Rchb.f.
Calanthe actinomorpha Fukuy.
Calanthe albo-longicalcarata S.S.Ying
Calanthe albolutea Ridl.
Calanthe alismaefolia Lindl.
Calanthe alpina Hook.f.
Calanthe angusta Lindl.
Calanthe angustifolia (Blume) Lindl.

CHINA (continued)

Calanthe arcuata Rolfe
Calanthe arcuata var. **brevifolia** Z.H.Tsi
Calanthe argenteostriata C.Z.Tang & S.J.Cheng
Calanthe arisanensis Hayata
Calanthe aristulifera Rchb.f.
Calanthe biloba Lindl.
Calanthe brevicornu Lindl.
Calanthe buccinifera Rolfe ex Hemsl.
Calanthe caudatilabella Hayata
Calanthe clavata Lindl.
Calanthe davidii Franch.
Calanthe delavayi Finet
Calanthe densiflora Lindl.
Calanthe discolor Lindl.
Calanthe discolor ssp. **discolor**
Calanthe dulongensis H.Li
Calanthe ecarinata Rolfe
Calanthe emeishanica Z.H.Tsi & K.Y.Lang
Calanthe fargesii Finet
Calanthe formosana Rolfe
Calanthe graciliflora Hayata
Calanthe griffithii Lindl.
Calanthe hancockii Rolfe
Calanthe henryi Rolfe
Calanthe herbacea Lindl.
Calanthe hirsuta Seidenf.
Calanthe labrosa (Rchb.f) Rchb.f.
Calanthe lechangensis Z.H.Tsi
Calanthe limprichtii Schltr.
Calanthe lyroglossa Rchb.f.
Calanthe mannii Hook.f.
Calanthe metoensis Z.H.Tsi
Calanthe nankunensis Z.H.Tsi
Calanthe nipponica Mak.
Calanthe odora Griff.
Calanthe petelotiana Gagnep.
Calanthe plantaginea Lindl.
Calanthe puberula Lindl.
Calanthe pumila Fukuy.
Calanthe reflexa Maxim.
Calanthe sacculata Schltr.
Calanthe simplex Seidenf.
Calanthe sinica Z.H.Tsi
Calanthe striata R.Br. ex Lindl.
Calanthe sylvatica (Thouars) Lindl.
Calanthe tangmaiensis K.Y.Lang & Tateishi
Calanthe tricarinata Lindl.
Calanthe trifida Tang & F.T.Wang
Calanthe triplicata (Willem.) Ames

CHINA (continued)

Calanthe tsoongiana Tang & F.T.Wang
Calanthe whiteana King & Pantl.
Calanthe yuana Tang & F.T.Wang
Renanthera coccinea Lour.
Renanthera imschootiana Rolfe
Rhynchostylis gigantea (Lindl.) Ridl.
Rhynchostylis retusa (L.) Blume
Vanda alpina (Lindl.) Lindl.
Vanda brunnea Rchb.f.
Vanda coerulea Griff. ex Lindl.
Vanda coerulescens Griff.
Vanda concolor Blume
Vanda cristata Lindl.
Vanda cruenta Lodd. Cat. ex Sweet
Vanda lamellata Lindl.
Vanda lilacina Teijsm. & Binn.
Vanda pumila Hook.f.
Vanda subconcolor Tang & F.T.Wang
Vanda taiwaniana S.S.Ying
Vandopsis gigantea (Lindl.) Pfitzer
Vandopsis undulata (Lindl.) J.J.Sm.

COLOMBIA / COLOMBIE (LA) / COLOMBIA

Brassavola cucullata (L.) R.Br.
Brassavola martiana Lindl.
Brassavola nodosa (L.) Lindl.
Brassavola venosa Lindl.
Calanthe calanthoides (A.Rich. & Galeottii) Hamer & Garay
Catasetum barbatum (Lindl.) Lindl.
Catasetum bergoldianum Foldats
Catasetum bicallosum Cogn.
Catasetum bicolor Klotzsch
Catasetum callosum Lindl.
Catasetum charlesworthii (Mansfield) Jenny
Catasetum collare Cogn.
Catasetum discolor (Lindl.) Lindl.
Catasetum fimbriatum (Morren) Lindl. & Paxton
?Catasetum lindleyanum Mansf.
Catasetum longifolium Lindl.
Catasetum lucis P.Ortiz & Arango
Catasetum macrocarpum Rich. ex Kunth
Catasetum microglossum Rolfe
Catasetum naso Lindl.
Catasetum ochraceum Lindl.
Catasetum pileatum Rchb.f.
Catasetum planiceps Lindl.
Catasetum × **roseo-album** (Hook.) Lindl.
Catasetum saccatum Lindl.

COLOMBIA (continued)

Catasetum sanguineum Lindl. & Paxton
Catasetum spinosum (Hook.) Lindl.
Catasetum tabulare Lindl.
Catasetum × **tapiriceps** Rchb.f.
Catasetum tricorne P.Ortiz
Catasetum tuberculatum Dodson
Miltoniopsis phalaenopsis (Rchb.f.) Garay & Dunst.
Miltoniopsis roezlii (Rchb.f.) God.-Leb.
Miltoniopsis santanaei Garay & Dunst.
Miltoniopsis vexillaria (Rchb.f) God.-Leb.
Renanthera striata Rolfe

COMOROS (THE) / COMORES (LES) / COMORAS (LAS)

Aerangis articulata (Rchb.f.) Schltr.
Aerangis modesta (Hook.f.) Schltr.
Aerangis mooreana (Rolfe ex Sander) P.J.Cribb & J.Stewart
Aerangis rostellaris (Rchb.f.) H.Perrier
Aerangis spiculata (Finet) Sengas
Aerangis stylosa (Rolfe) Schltr.
Angraecum calceolus Thouars
Angraecum eburneum Bory
Angraecum eburneum ssp. **superbum** Thouars
Angraecum florulentum Rchb.f.
Angraecum germinyanum Hook.f.
Angraecum leonis (Rchb.f.) Andre
Angraecum meirax (Rchb.f.) H.Perrier
Angraecum pectinatum Thouars
Angraecum scottianum Rchb.f.
Angraecum vesiculatum Schltr.
Angraecum vesiculiferum Schltr.
Angraecum xylopus Rchb.f.
Calanthe sylvatica (Thouars) Lindl.

CONGO (THE) / CONGO (LE) / CONGO (EL)

Angraecum affine Schltr.
Angraecum bancoense Van der Burg
Angraecum distichum Lindl.
Angraecum moandense De Wild.

CONGO (THE DEMOCRATIC REPUBLIC OF THE) / RÉPUBLIQUE DÉMOCRATIQUE DU CONGO (LA) / REPÚBLICA DEMOCRÁTICA DEL CONGO (LA)

Aerangis arachnopus (Rchb.f.) Schltr.
Aerangis calantha (Schltr.) Schltr.
Aerangis collum-cygni Summerh.
Aerangis kotschyana (Rchb.f.) Schltr.

CONGO (THE DEMOCRATIC REPUBLIC OF THE) (Continued)

Aerangis luteo-alba (Kraenzl.) Schltr.
Aerangis luteo-alba var. **luteo-alba**
Aerangis luteo-alba var. **rhodosticta** (Kraenzl.) J.Stewart
Aerangis stelligera Summerh.
Aerangis ugandensis Summerh.
Aerangis verdickii (De Wild.) Schltr.
Angraecum affine Schltr.
Angraecum angustipetalum Rendle
Angraecum aporoides Summerh.
Angraecum claessensii De Wild.
Angraecum distichum Lindl.
Angraecum eichlerianum Kraenzl.
Angraecum gabonense Summerh.
Angraecum infundibulare Lindl.
Angraecum mofakoko De Wild.
Angraecum podochiloides Schltr.
Angraecum pungens Schltr.
Angraecum reygaertii De Wild.
Angraecum sacciferum Lindl.
Angraecum stolzii Schltr.
Angraecum subulatum Lindl.
Calanthe sylvatica (Thouars) Lindl.

COSTA RICA / COSTA RICA / COSTA RICA (LE) / COSTA RICA

Brassavola acaulis Lindl.
Brassavola grandiflora Lindl.
Brassavola nodosa (L.) Lindl.
Calanthe calanthoides (A.Rich. & Galeottii) Hamer & Garay
Catasetum integerrimum Hook.
Catasetum maculatum Kunth
Catasetum sanguineum Lindl. & Paxton
Catasetum viridiflavum Hook.
Miltonioides carinifera (Rchb.f.) Senghas & Luckel
Miltonioides leucomelas (Rchb.f.) Bockemuhl & Senghas
Miltonioides schroederiana (O'Brien) Brieger & Luckel
Miltoniopsis vexillaria (Rchb.f) God.-Leb.
Miltioniopsis warscewiczii (Rchb.f.) Garay & Dunst.
Rossioglossum powellii (Schltr.) Garay & G.C.Kenn.
Rossioglossum schlieperianum (Rchb.f.) Garay & G.C.Kenn.
Rossioglossum schlieperianum var. **flavidum** (Rchb.f.) Garay & G.C.Kenn.
Rossioglossum williamsianum (Rchb.f.) Garay & G.C.Kenn.

CÔTE D'IVOIRE / CÔTE D'IVOIRE (LA) / CÔTE D'IVOIRE (LA)

Aerangis biloba (Lindl.) Schltr.
Angraecum bancoense Van der Burg
Angraecum birrimense Rolfe
Angraecum distichum Lindl.

CÔTE D'IVOIRE (Continued)

Angraecum moandense De Wild.
Angraecum podochiloides Schltr.
Angraecum pyriforme Summerh.
Angraecum subulatum Lindl.

CUBA / CUBA / CUBA

Calanthe calanthoides (A.Rich. & Galeottii) Hamer & Garay

DOMINICAN REPUBLIC (THE) / RÉPUBLIQUE DOMINICAINE (LA) / REPÚBLICA DOMINICANA (LA)

Calanthe calanthoides (A.Rich. & Galeottii) Hamer & Garay

ECUADOR / EQUATEUR (L') / ECUADOR (EL)

Brassavola nodosa (L.) Lindl.
Catasetum barbatum (Lindl.) Lindl.
Catasetum expansum Rchb.f.
Catasetum fimbriatum (Morren) Lindl. & Paxton
?Catasetum lindleyanum Mansf.
Catasetum macroglossum Rchb.f.
Catasetum microglossum Rolfe
Catasetum napoense Dodson
Catasetum saccatum Lindl.
Catasetum samaniegoi Dodson
Catasetum × **sodiroi** Schltr.
Catasetum stevensonii Dodson
Catasetum tuberculatum Dodson
Miltoniopsis roezlii (Rchb.f.) God.-Leb.
Miltoniopsis santanaei Garay & Dunst.
Miltoniopsis vexillaria (Rchb.f.) God.-Leb.
Renanthera striata Rolfe

EL SALVADOR / EL SALVADOR / EL SALVADOR

Brassavola cucullata (L.) R.Br.
Brassavola grandiflora Lindl.
Brassavola nodosa (L.) Lindl.
Catasetum integerrimum Hook.
Miltonioides leucomelas (Rchb.f.) Bockemuhl & Senghas

EQUATORIAL GUINEA / GUINÉE ÉQUATORIALE (LA) / GUINEA ECUATORIAL (LA)

Aerangis calantha (Schltr.) Schltr.
Aerangis gravenreuthii (Kraenzl.) Schltr.
Aerangis megaphylla Summerh.
Angraecum affine Schltr.

EQUATORIAL GUINEA (Continued)

Angraecum aporoides Summerh.
Angraecum lisowskianum Szl. & Olzs.
Angraecum moandense De Wild.
Angraecum pungens Schltr.
Angraecum subulatum Lindl.
Calanthe sylvatica (Thouars) Lindl.

ETHIOPIA / ETHIOPIE (L') / ETIOPIA

Aerangis brachycarpa (A.Rich.) Dur. & Schinz
Aerangis kotschyana (Rchb.f.) Schltr.
Aerangis luteo-alba (Kraenzl.) Schltr.
Aerangis luteo-alba var. **rhodosticta** (Kraenzl.) J.Stewart
Aerangis somalensis (Schltr.) Schltr.
Angraecum infundibulare Lindl.
Angraecum minus Summerh.

FIJI / FIDJI (LES) / FIJI

Calanthe alta Rchb.f.
Calanthe hololeuca Rchb.f.
Calanthe imthurnii Kores
?Calanthe triplicata (Willem.) Ames
Calanthe ventilabrum Rchb.f.

FRENCH GUIANA / GUYANE FRANCAISE / GUYANA FRANCESCA

Brassavola martiana Lindl.
Catasetum longifolium Lindl.
Catasetum macrocarpum Rich. ex Kunth

FRENCH POLYNESIA / POLYNESIE FRANCAISE / POLINESIA FRANCESCA

Calanthe triantherifera Nadeaud

GABON / GABON (LE) / GABÓN (EL)

Aerangis arachnopus (Rchb.f.) Schltr.
Aerangis gracillima (Kraenzl.) J.C.Arends & J.Stewart
Angraecum angustipetalum Rendle
Angraecum cribbianum Szl. & Olzs.
Angraecum distichum Lindl.
Angraecum egertonii Rendle
Angraecum eichlerianum Kraenzl.
Angraecum eichlerianum var. **curvicalcaratum** Szl. & Olzs.
Angraecum gabonense Summerh.
Angraecum moandense De Wild.
Angraecum multinominatum Rendle
Calanthe sylvatica (Thouars) Lindl.

GHANA / GHANA (LE) / GHANA

Aerangis arachnopus (Rchb.f.) Schltr.
Aerangis biloba (Lindl.) Schltr.
Aerangis calantha (Schltr.) Schltr.
Angraecum angustipetalum Rendle
Angraecum birrimense Rolfe
Angraecum distichum Lindl.
Angraecum multinominatum Rendle
Angraecum podochiloides Schltr.
Angraecum subulatum Lindl.

GUATEMALA / GUATEMALA (LE) / GUATEMALA

Brassavola acaulis Lindl.
Brassavola cucullata (L.) R.Br.
Brassavola grandiflora Lindl.
Brassavola nodosa (L.) Lindl.
Calanthe calanthoides (A.Rich. & Galeottii) Hamer & Garay
Catasetum integerrimum Hook.
Miltonioides laevis (Lindl.) Brieger & Luckel
Miltonioides leucomelas (Rchb.f.) Bockemuhl & Senghas
Rossioglossum grande (Lindl.) Garay & G.C.Kenn.
Rossioglossum grande var. **aureum** (Hort. ex Stein) Garay & G.C.Kenn.
Rossioglossum williamsianum (Rchb.f.) Garay & G.C.Kenn.

GUINEA / GUINÉE (LA) / GUINEA

Aerangis biloba (Lindl.) Schltr.
Aerangis kotschyana (Rchb.f.) Schltr.
Angraecum distichum Lindl.
Angraecum moandense De Wild.
Angraecum multinominatum Rendle
Angraecum nzoanum A.Chev.
Calanthe sylvatica (Thouars) Lindl.

GUYANA / GUYANA (LE) / GUYANA

Brassavola angustata Lindl.
Brassavola cucullata (L.) R.Br.
Brassavola gardneri Cogn.
Brassavola martiana Lindl.
Brassavola nodosa (L.) Lindl.
Catasetum barbatum (Lindl.) Lindl.
Catasetum cornutum Lindl.
Catasetum × **guianense** G.A.Romero & Jenny
Catasetum longifolium Lindl.
Catasetum macrocarpum Rich. ex Kunth
Catasetum planiceps Lindl.
Catasetum poriferum Lindl.
Catasetum saccatum Lindl.

GUYANA (Continued)

Catasetum spinosum (Hook.) Lindl.
Catasetum thompsonii Dodson

HAITI / HAÏTI / HAITÍ

Calanthe calanthoides (A.Rich. & Galeottii) Hamer & Garay

HONDURAS / HONDURAS (LE) / HONDURAS

Brassavola cucullata (L.) R.Br.
Brassavola grandiflora Lindl.
Brassavola nodosa (L.) Lindl.
Calanthe calanthoides (A.Rich. & Galeottii) Hamer & Garay
Catasetum integerrimum Hook.
Miltonioides leucomelas (Rchb.f.) Bockemuhl & Senghas
Rossioglossum williamsianum (Rchb.f.) Garay & G.C.Kenn.

INDIA / INDE (L') / INDIA (LA)

Ascocentrum ampullaceum (Roxb.) Schltr.
Ascocentrum curvifolium (Lindl.) Schltr.
Ascocentrum himalaicum (Deb., Sengupta & Malick) Christenson
Ascocentrum semiteretifolium Seidenf.
Calanthe alismaefolia Lindl.
Calanthe alpina Hook.f.
Calanthe angusta Lindl.
Calanthe anjanae Lucksom
Calanthe biloba Lindl.
Calanthe brevicornu Lindl.
Calanthe clavata Lindl.
Calanthe fulgens Lindl.
Calanthe griffithii Lindl.
Calanthe herbacea Lindl.
Calanthe mannii Hook.f.
Calanthe pachystalix Rchb.f.
Calanthe plantaginea Lindl.
Calanthe puberula Lindl.
Calanthe pulchra (Blume) Lindl.
Calanthe sylvatica (Thouars) Lindl.
Calanthe tricarinata Lindl.
Calanthe triplicata (Willem.) Ames
Calanthe uncata Lindl.
Calanthe whiteana King & Pantl.
Calanthe yoksomnensis Lucksom
Renanthera imschootiana Rolfe
Rhynchostylis retusa (L.) Blume
Vanda alpina (Lindl.) Lindl.
Vanda arbuthnotiana Kraenzl.
Vanda coerulea Griff. ex Lindl.

INDIA (Continued)

Vanda coerulescens Griff.
Vanda cristata Lindl.
Vanda jainii A.S.Chauhan
Vanda pumila Hook.f.
Vanda spathulata (L.) Spreng.
Vanda stangeana Rchb.f.
Vanda tessellata (Roxb.) Hook. ex G.Don
Vanda testacea (Lindl.) Rchb.f.
Vanda wightii Rchb.f.
Vandopsis undulata (Lindl.) J.J.Sm.

INDONESIA / INDONÉSIE (L') / INDONESIA

Ascocentrum aurantiacum (Schltr.) Schltr.
Ascocentrum aureum J.J.Sm.
Ascocentrum hendersoniana Schltr.
Ascocentrum insularum Christenson
Ascocentrum miniatum (Lindl.) Schltr.
Calanthe × **albolilacina** J.J.Sm.
Calanthe × **subhamata** J.J.Sm.
Calanthe × **varians** J.J.Sm.
Calanthe abbreviata (Blume) Lindl.
Calanthe albolutea Ridl.
Calanthe angustifolia (Blume) Lindl.
Calanthe arfakana J.J.Sm.
Calanthe atjehensis J.J.Sm.
Calanthe aurantiaca Ridl.
Calanthe aurantimacula P.van Royen
Calanthe aureiflora J.J.Sm.
Calanthe baliensis J.J.Wood & J.B.Comber
Calanthe bicalcarata J.J.Sm.
Calanthe breviflos Ridl.
Calanthe caulescens J.J.Sm.
Calanthe caulodes J.J.Sm.
Calanthe ceciliae Rchb.f.
Calanthe chrysoglossoides J.J.Sm.
Calanthe clavicalcar J.J.Sm.
Calanthe crenulata J.J.Sm.
Calanthe crumenata Ridl.
Calanthe dipteryx Rchb.f.
Calanthe ecallosa J.J.Sm.
Calanthe engleriana Kraenzl.
Calanthe epiphytica Carr
Calanthe flava (Blume) Morren
Calanthe forbesii Ridl.
Calanthe geelvinkensis J.J.Sm.
Calanthe gibbsiae Rolfe
Calanthe graciliscapa Schltr.
Calanthe hyacinthina Schltr.

INDONESIA (Continued)

Calanthe kemulense J.J.Sm.
Calanthe leucosceptrum Schltr.
Calanthe longibracteata Ridl.
Calanthe melinosema Schltr.
Calanthe microglossa Ridl.
Calanthe moluccensis J.J.Sm.
Calanthe nicolae P.O'Byrne
Calanthe obreniformis J.J.Sm.
Calanthe pauciverrucosa J.J.Sm.
Calanthe pulchra (Blume) Lindl.
Calanthe pullei J.J.Sm.
Calanthe rajana J.J.Sm.
Calanthe reflexilabris J.J.Sm.
Calanthe rhodochila Schltr.
Calanthe rhodochila var. **reconditiflora** (J.J.Sm.) S.Thomas
Calanthe rigida Carr
Calanthe ruttenii J.J.Sm.
Calanthe saccata J.J.Sm.
Calanthe salaccensis J.J.Sm.
Calanthe seranica J.J.Sm.
Calanthe shelfordii Ridl.
Calanthe speciosa (Blume) Lindl.
Calanthe sylvatica (Thouars) Lindl.
Calanthe taenioides J.J.Sm.
Calanthe transiens J.J.Sm.
Calanthe triplicata (Willem.) Ames
Calanthe truncata J.J.Sm.
Calanthe truncicola Schltr.
Calanthe undulata J.J.Sm.
Calanthe unifolia Ridl.
Calanthe ventilabrum Rchb.f.
Calanthe versteegii J.J.Sm.
Calanthe vestita Lindl.
Calanthe villosa J.J.Sm.
Calanthe zollingeri Rchb.f.
Renanthera edefeldtii F.Muell. & Kraenzl. ex Kraenzl.
Renanthera elongata (Blume) Lindl.
Renanthera isosepala Holttum
Renanthera matutina Lindl.
Renanthera moluccana Blume
Renanthera sulingi Lindl.
Rhynchostylis gigantea (Lindl.) Ridl.
Rhynchostylis retusa (L.) Blume
Vanda arcuata J.J.Sm.
Vanda × **boumaniae** J. J. Smith
Vanda celebica Rolfe
Vanda crassiloba Teijsm. & Binn.
Vanda dearei Rchb.f.
Vanda devoogtei J.J.Sm.

INDONESIA (Continued)

Vanda foetida J.J.Sm.
Vanda hastifera Rchb.f.
Vanda helvola Blume
Vanda hindsii Lindl.
Vanda insignis Blume
Vanda lamellata Lindl.
Vanda leucostele Schltr.
Vanda limbata Blume
Vanda lindeni Rchb.f.
Vanda liouvillei Finet
Vanda lombokensis J.J.Sm.
Vanda luzonica Loher ex Rolfe
Vanda moorei Rolfe
Vanda pauciflora Breda
Vanda pumila Hook.f.
Vanda punctata Ridl.
Vanda scandens Holttum
Vanda sumatrana Schltr.
Vanda tricolor Lindl.
Vanda tricolor ssp. **suavis** (Lindl.) Veitch
Vandopsis lissochiloides (Gaudich.) Pfitzer

JAMAICA / JAMAÏQUE (LA / JAMAICA)

Calanthe calanthoides (A.Rich. & Galeottii) Hamer & Garay

JAPAN / JAPON (LE) / JAPÓN (EL)

Angraecum pygmaecum Linden
Bletilla striata (Thunb. ex A.Murray) Rchb.f.
Calanthe alismaefolia Lindl.
Calanthe alpina Hook.f.
Calanthe aristulifera Rchb.f.
Calanthe davidii Franch.
Calanthe densiflora Lindl.
Calanthe discolor ssp. **amamiana** (Fukuy.) Masam.
Calanthe discolor ssp. **discolor**
Calanthe discolor forma **quinquelamellata** Hiroe
Calanthe discolor ssp. **kanashiroi** Fukuy.
Calanthe discolor ssp. **tokunoshimensis** (Hatus. & Ida) Hatus.
Calanthe formosana Rolfe
Calanthe hattorii Schltr.
Calanthe hoshii S.Kobay.
Calanthe izu-insularis (Satomi) Ohwi. & Satomi
Calanthe laxiflora Makino
Calanthe lyroglossa Rchb.f.
Calanthe mannii Hook.f.
Calanthe nipponica Mak.
Calanthe × **oodaruma** Maekawa

JAPAN (Continued)

Calanthe reflexa Maxim.
Calanthe striata R.Br. ex Lindl
Calanthe sylvatica (Thouars) Lindl.
Calanthe tricarinata Lindl.
Calanthe triplicata (Willem.) Ames
Calanthe tyoh-harai Makino
Vanda lamellata Lindl.

KENYA / KENYA (LE) / KENYA

Aerangis brachycarpa (A.Rich.) Dur. & Schinz
Aerangis confusa J.Stewart
Aerangis coriacea Summerh.
Aerangis hologlottis (Schltr.) Schltr.
Aerangis kirkii (Rchb.f.) Schltr.
Aerangis kotschyana (Rchb.f.) Schltr.
Aerangis luteo-alba (Kraenzl.) Schltr.
Aerangis luteo-alba var. **rhodosticta** (Kraenzl.) J.Stewart
Aerangis somalensis (Schltr.) Schltr.
Aerangis thomsonii (Rolfe) Schltr.
Aerangis ugandensis Summerh.
Angraecum chamaeanthus Schltr.
Angraecum conchiferum Lindl.
Angraecum cultriforme Summerh.
Angraecum decipiens Summerh.
Angraecum dives Rolfe
Angraecum eburneum ssp. **giryamae** (Rendle) Senghas & Cribb
Angraecum erectum Summerh.
Angraecum firthii Summerh.
Angraecum humile Summerh.
Angraecum infundibulare Lindl.
Angraecum keniae Kraenzl.
Angraecum sacciferum Lindl.
Angraecum teres Summerh.
Angraecum viride Kraenzl.
Calanthe sylvatica (Thouars) Lindl.

KOREA (THE DEMOCRATIC PEOPLE'S REPUBLIC OF) / RÉPUBLIQUE POPULAIRE DÉMOCRATIQUE DE CORÉE (LA) / REPÚBLICA POPULAR DEMOCRÁTICA DE COREA (LA)

Bletilla striata (Thunb. ex A.Murray) Rchb.f.
Calanthe discolor ssp. **discolor**

KOREA (THE REPUBLIC OF) / RÉPUBLIQUE DE CORÉE (LA) / REPÚBLICA DE COREA (LA)

Bletilla striata (Thunb. ex A.Murray) Rchb.f.
Calanthe coreana Nakai

KOREA (THE REPUBLIC OF) (Continued)

Calanthe discolor ssp. **discolor**
Calanthe reflexa Maxim.

LAO PEOPLE'S DEMOCRATIC REPUBLIC (THE) / RÉPUBLIQUE DÉMOCRATIQUE POPULAIRE LAO (LA) / REPÚBLICA DEMOCRÁTICA POPULAR LAO (LA)

Ascocentrum ampullaceum (Roxb.) Schltr.
Ascocentrum garayi Christenson
Ascocentrum pusillum Aver.
Calanthe angusta Lindl.
Calanthe cardioglossa Schltr.
Calanthe lyroglossa Rchb.f.
Calanthe poilanei Gagnep.
Calanthe succedanea Gagnep.
Calanthe triplicata (Willem.) Ames
Renanthera coccinea Lour.
Rhynchostylis coelestis Rchb.f.
Rhynchostylis gigantea (Lindl.) Ridl.
Rhynchostylis retusa (L.) Blume
Vanda concolor Blume
Vanda denisoniana Benson & Rchb.f.
Vanda lilacina Teijsm. & Binn.
Vanda liouvillei Finet
Vanda pumila Hook.f.
Vanda tricolor ssp. **suavis** (Lindl.) Veitch
Vandopsis gigantea (Lindl.) Pfitzer
Vandopsis lissochiloides (Gaudich.) Pfitzer

LIBERIA / LIBÉRIA / LIBERIA

Aerangis biloba (Lindl.) Schltr.
Angraecum birrimense Rolfe
Angraecum claessensii De Wild.
Angraecum distichum Lindl.
Angraecum moandense De Wild.
Angraecum modicum Summerh.
Angraecum podochiloides Schltr.

MADAGASCAR / MADAGASCAR / MADAGASCAR

Aerangis articulata (Rchb.f.) Schltr.
Aerangis citrata (Thouars) Schltr.
Aerangis concavipetala H.Perrier
Aerangis cryptodon (Rchb.f.) Schltr.
Aerangis curnowiana (Finet) H.Perrier
Aerangis decaryana H.Perrier
Aerangis ellisii (B.S.Williams) Schltr.
Aerangis ellisii ssp. **grandiflora** J.Stewart

MADAGASCAR (Continued)

Aerangis fastuosa (Rchb.f.) Schltr.
Aerangis fuscata (Rchb.f.) Schltr.
Aerangis hyaloides (Rchb.f.) Schltr.
Aerangis macrocentra (Schltr.) Schltr.
Aerangis modesta (Hook.f.) Schltr.
Aerangis mooreana (Rolfe ex Sander) P.J.Cribb & J.Stewart
Aerangis pallida (W.Watson) Garay
Aerangis pallidiflora H.Perrier
Aerangis × **primulina** (Rolfe) H.Perrier
Aerangis pulchella (Schltr.) Schltr.
Aerangis punctata J.Stewart
Aerangis rostellaris (Rchb.f.) H.Perrier
Aerangis spiculata (Finet) Sengas
Aerangis stylosa (Rolfe) Schltr.
Angraecum acutipetalum ssp. **analabeensis** H.Perrier
Angraecum acutipetalum ssp. **ankeranae** H.Perrier
Angraecum acutipetalum Schltr.
Angraecum alleizettei Schltr.
Angraecum aloifolium Hermans & P.J.Cribb
Angraecum ambrense H.Perrier
Angraecum amplexicaule Toill.-Gen. & Bosser
Angraecum ampullaceum Bosser
Angraecum andasibeense H.Perrier
Angraecum andringitranum Schltr.
Angraecum ankeranense H.Perrier
Angraecum appendiculoides Schltr.
Angraecum aviceps Schltr.
Angraecum baronii (Finet) Schltr.
Angraecum bemarivoense Schltr.
Angraecum bicallosum H.Perrier
Angraecum brachyrhopalon Schltr.
Angraecum breve Schltr.
Angraecum calceolus Thouars
Angraecum caricifolium H.Perrier
Angraecum caulescens Thouars
Angraecum chaetopodum Schltr.
Angraecum chermezoni H.Perrier
Angraecum chloranthum Schltr.
Angraecum clavigerum Ridl.
Angraecum compactum Schltr.
Angraecum compressicaule H.Perrier
Angraecum coriaceum (Thunb. ex Sw.) Schltr.
Angraecum cornucopiae H.Perrier
Angraecum corynoceras Schltr.
Angraecum coutrixii Bosser
Angraecum crassum Thouars
Angraecum curnowianum (Rchb.f.) Dur. & Schinz
Angraecum curvicalcar Schltr.
Angraecum curvicaule Schltr.

MADAGASCAR (Continued)

Angraecum danguyanum H.Perrier
Angraecum dasycarpum Schltr.
Angraecum dauphinense (Rolfe) Schltr.
Angraecum decaryanum H.Perrier
Angraecum dendrobiopsis Schltr.
Angraecum didieri (Baill. ex Finet) Schltr.
Angraecum dollii Senghas
Angraecum drouhardii H.Perrier
Angraecum dryadum Schltr.
Angraecum eburneum Bory
Angraecum eburneum ssp. **superbum** Thouars
Angraecum eburneum var. **longicalcar** Bosser
Angraecum eburneum ssp. **xerophilum** H.Perrier
Angraecum elephantinum Schltr.
Angraecum elliotii Rolfe
Angraecum equitans Schltr.
Angraecum falcifolium Bosser
Angraecum ferkoanum Schltr.
Angraecum filicornu Thouars
Angraecum flavidum Bosser
Angraecum floribundum Bosser
Angraecum florulentum Rchb.f.
Angraecum germinyanum Hook.f.
Angraecum guillauminii H.Perrier
Angraecum humbertii H.Perrier
Angraecum humblotianum (Finet) Schltr.
Angraecum huntleyoides Schltr.
Angraecum imerinense Schltr.
Angraecum implicatum Thouars
Angraecum inapertum Thouars
Angraecum kraenzlinianum H.Perrier
Angraecum laggiarae Schltr.
Angraecum lecomtei H.Perrier
Angraecum leonis (Rchb.f.) Andre
Angraecum letouzeyi Bosser
Angraecum linearifolium Garay
Angraecum litorale Schltr.
Angraecum longicaule H.Perrier
Angraecum madagascariense (Finet) Schltr.
Angraecum magdalenae Schltr.
Angraecum magdalenae var. **latilabellum** Bosser
Angraecum mahavavense H.Perrier
Angraecum mauritianum (Poir.) Frapp.
Angraecum meirax (Rchb.f.) H.Perrier
Angraecum melanostictum Schltr.
Angraecum microcharis Schltr.
Angraecum mirabile Schltr.
Angraecum moratii Bosser
Angraecum multiflorum Thouars

MADAGASCAR (Continued)

Angraecum muscicolum H.Perrier
Angraecum musculiferum H.Perrier
Angraecum myrianthum Schltr.
Angraecum nasutum Schltr.
Angraecum obesum H.Perrier
Angraecum oblongifolium Toill.-Gen. & Bosser
Angraecum ochraceum (Ridl.) Schltr.
Angraecum onivense H.Perrier
Angraecum palmicolum Bosser
Angraecum panicifolium H.Perrier
Angraecum pauciramosum Schltr.
Angraecum pectinatum Thouars
Angraecum penzigianum Schltr.
Angraecum pergracile Schltr.
Angraecum perhumile H.Perrier
Angraecum perparvulum H.Perrier
Angraecum peyrotii Bosser
Angraecum pinifolium Bosser
Angraecum popowii Braem
Angraecum potamophilum Schltr.
Angraecum praestans Schltr.
Angraecum protensum Schltr.
Angraecum pseudodidieri H.Perrier
Angraecum pseudofilicornu H.Perrier
Angraecum pterophyllum H.Perrier
Angraecum pumilio Schltr.
Angraecum ramulicolum H.Perrier
Angraecum rhizanthium H.Perrier
Angraecum rhizomaniacum Schltr.
Angraecum rhynchoglossum Schltr.
Angraecum rigidifolium H.Perrier
Angraecum rostratum Ridl.
Angraecum rubellum Bosser
Angraecum rutenbergianum Kraenzl.
Angraecum sacculatum Schltr.
Angraecum sambiranoense Schltr.
Angraecum scalariforme H.Perrier
Angraecum sedifolium Schltr.
Angraecum serpens (H.Perrier) Bosser
Angraecum sesquipedale Thouars
Angraecum sesquipedale ssp. **angustifolium** Bosser & Morat
Angraecum sesquisectangulum Kraenzl.
Angraecum setipes Schltr.
Angraecum sinuatiflorum H.Perrier
Angraecum sororium Schltr.
Angraecum sterrophyllum Schltr.
Angraecum tamarindicolum Schltr.
Angraecum tenellum (Ridl.) Schltr.
Angraecum tenuipes Summerh.

MADAGASCAR (Continued)

Angraecum tenuispica Schltr.
Angraecum teretifolium Ridl.
Angraecum triangulifolium Senghas
Angraecum trichoplectron (Rchb.f.) Schltr.
Angraecum urschianum Toill.-Gen. & Bosser
Angraecum verecundum Schltr.
Angraecum vesiculatum Schltr.
Angraecum vesiculiferum Schltr.
Angraecum viguieri Schltr.
Angraecum zaratananae Schltr.
Calanthe humbertii H.Perrier
Calanthe madagascariensis Rolfe
Calanthe millotae Ursch & Toill.-Gen. ex Bosser
Calanthe repens Schltr.
Calanthe sylvatica (Thouars) Lindl.

MALAWI / MALAWI (LE) / MALAWI

Aerangis alcicornis (Rchb.f.) Garay
Aerangis appendiculata (De Wild.) Schltr.
Aerangis carnea J.Stewart
Aerangis distincta J.Stewart & la Croix
Aerangis kirkii (Rchb.f.) Schltr.
Aerangis kotschyana (Rchb.f.) Schltr.
Aerangis montana J.Stewart
Aerangis mystacidii (Rchb.f.) Schltr.
Aerangis oligantha Schltr.
Aerangis somalensis (Schltr.) Schltr.
Aerangis splendida J.Stewart & la Croix
Aerangis verdickii (De Wild.) Schltr.
Angraecum angustipetalum Rendle
Angraecum chamaeanthus Schltr.
Angraecum conchiferum Lindl.
Angraecum cultriforme Summerh.
Angraecum sacciferum Lindl.
Angraecum stella-africae P.J.Cribb
Angraecum stolzii Schltr.
Angraecum umbrosum P.J.Cribb
Calanthe sylvatica (Thouars) Lindl.

MALAYSIA / MALAISIE (LA) / MALASIA

Ascocentrum garayi Christenson
Ascocentrum hendersoniana Schltr.
Calanthe albolutea Ridl.
Calanthe angustifolia (Blume) Lindl.
Calanthe aurantiaca Ridl.
Calanthe aureiflora J.J.Sm.
Calanthe carrii Seidenf. & J.J.Wood nom. nov.

MALAYSIA (Continued)

Calanthe ceciliae Rchb.f.
Calanthe clavata Lindl.
Calanthe cleistogama Holttum
Calanthe conspicua Lindl.
Calanthe crenulata J.J.Sm.
Calanthe gibbsiae Rolfe
Calanthe johorensis Holttum
Calanthe kinabaluensis Rolfe
Calanthe lyroglossa Rchb.f.
Calanthe monophylla Ridl.
Calanthe otuhanica C.L.Chan & T.J.Barkman
Calanthe ovalifolia Ridl.
Calanthe ovata Ridl.
Calanthe pubescens Ridl.
Calanthe pulchra (Blume) Lindl.
Calanthe rigida Carr
Calanthe rubens Ridl.
Calanthe salaccensis J.J.Sm.
Calanthe speciosa (Blume) Lindl.
Calanthe sylvatica (Thouars) Lindl.
Calanthe taenioides J.J.Sm.
Calanthe tenuis Ames & C.Schweinf.
Calanthe transiens J.J.Sm.
Calanthe triplicata (Willem.) Ames
Calanthe truncicola Schltr.
Calanthe undulata J.J.Sm.
Calanthe vestita Lindl.
Renanthera bella J.J.Wood
Renanthera elongata (Blume) Lindl.
Renanthera isosepala Holttum
Renantherella histrionica (Rchb.f.) Ridl.
Rhynchostylis gigantea (Lindl.) Ridl.
Rhynchostylis retusa (L.) Blume
Vanda dearei Rchb.f.
Vanda flavobrunnea Rchb.f.
Vanda hastifera Rchb.f.
Vanda helvola Blume
Vanda insignis Blume
Vanda lamellata Lindl.
Vanda liouvillei Finet
Vanda moorei Rolfe
Vanda punctata Ridl.
Vanda scandens Holttum
Vandopsis gigantea (Lindl.) Pfitzer

MAURITIUS / MAURICE / MAURICIO

Angraecum borbonicum Bosser
Angraecum cadetii Bosser

MAURITIUS (Continued)

Angraecum calceolus Thouars
Angraecum caulescens Thouars
Angraecum costatum Frapp. ex Cordem.
Angraecum divaricatum Frapp.
Angraecum eburneum Bory
Angraecum filicornu Thouars
Angraecum inapertum Thouars
Angraecum mauritianum (Poir.) Frapp.
Angraecum multiflorum Thouars
Angraecum nanum Frapp. ex Cordem.
Angraecum oberonia Finet
Angraecum obversifolium Frapp.
Angraecum palmiforme Thouars
Angraecum parvulum Ayres ex Baker
Angraecum pectinatum Thouars
Angraecum pingue Frapp.
Angraecum ramosum Thouars
Angraecum triquetrum Thouars
Angraecum undulatum (Cordem.) Schltr.
Angraecum yuccaefolium Bojer
Calanthe candida Bosser
Calanthe sylvatica (Thouars) Lindl.
Calanthe triplicata (Willem.) Ames

MEXICO / MEXIQUE (LE) / MÉXICO

Brassavola acaulis Lindl.
Brassavola cucullata (L.) R.Br.
Brassavola grandiflora Lindl.
Brassavola nodosa (L.) Lindl.
Calanthe calanthoides (A.Rich. & Galeottii) Hamer & Garay
Catasetum integerrimum Hook.
Catasetum laminatum Lindl.
Catasetum pendulum Dodson
Miltonioides karwinskii (Lindl.) Brieger & Luckel
Miltonioides laevis (Lindl.) Brieger & Luckel
Miltonioides leucomelas (Rchb.f.) Bockemuhl & Senghas
Miltonioides oviedomotae (Hagsater) Senghas
Miltonioides reichenheimii (Linden & Rchb.f.) Brieger & Luckel
Rossioglossum grande (Lindl.) Garay & G.C.Kenn.
Rossioglossum grande var. **aureum** (Hort. ex Stein) Garay & G.C.Kenn.
Rossioglossum insleayi (Barker ex Lindl.) Garay & G.C.Kenn.
Rossioglossum splendens (Rchb.f.) Garay & G.C.Kenn.
Rossioglossum splendens var. **imschootianum** (Rolfe) Garay & G.C.Kenn.
Rossioglossum splendens var. **leopardinum** (Regel) Garay & G.C.Kenn.
Rossioglossum splendens var. **pantherinum** (Rchb.f.) Garay & G.C.Kenn.

MICRONESIA (FEDERATED STATES OF) / MICRONÉSIE (ETATS FÉDÉRÉS DE) / MICRONESIA (ESTADOS FEDERADOS DE)

Calanthe triplicata (Willem.) Ames

MOZAMBIQUE / MOZAMBIQUE (LE) / MOZAMBIQUE

Aerangis alcicornis (Rchb.f.) Garay
Aerangis appendiculata (De Wild.) Schltr.
Aerangis hologlottis (Schltr.) Schltr.
Aerangis kirkii (Rchb.f.) Schltr.
Aerangis kotschyana (Rchb.f.) Schltr.
Aerangis mystacidii (Rchb.f.) Schltr.
Aerangis verdickii (De Wild.) Schltr.
Angraecum calceolus Thouars
Angraecum chamaeanthus Schltr.
Angraecum conchiferum Lindl.
Angraecum cultriforme Summerh.
Angraecum sacciferum Lindl.
Angraecum stolzii Schltr.

MYANMAR / MYANMAR (LE) / MYANMAR

Ascocentrum ampullaceum (Roxb.) Schltr.
Ascocentrum ampullaceum var. **auranticum** Pradhan
Ascocentrum curvifolium (Lindl.) Schltr.
Ascocentrum himalaicum (Deb., Sengupta & Malick) Christenson
Ascocentrum rubrum (Lindl.) Seidenf.
Bletilla chartacea (King & Pantl.) Tang & Wang
Bletilla foliosa (King & Pantl.) Tang & Wang
Calanthe biloba Lindl.
Calanthe ceciliae Rchb.f.
Calanthe clavata Lindl.
Calanthe densiflora Lindl.
Calanthe griffithii Lindl.
Calanthe labrosa (Rchb.f) Rchb.f.
Calanthe lyroglossa Rchb.f.
Calanthe rosea (Lindl.) Benth.
Calanthe triplicata (Willem.) Ames
Calanthe vestita Lindl.
Renanthera annamensis Rolfe
Renanthera coccinea Lour.
Renanthera imschootiana Rolfe
Rhynchostylis gigantea (Lindl.) Ridl.
Rhynchostylis retusa (L.) Blume
Vanda bensonii Bateman
Vanda brunnea Rchb.f.
Vanda × **charlesworthtii** Rolfe
Vanda coerulea Griff. ex Lindl.
Vanda coerulescens Griff.
?Vanda coerulescens ssp. **boxallii** Rchb.f.

MYANMAR (Continued)

Vanda × confusa Rolfe
Vanda cristata Lindl.
Vanda denisoniana Benson & Rchb.f.
Vanda lilacina Teijsm. & Binn.
Vanda liouvillei Finet
Vanda moorei Rolfe
Vanda tessellata (Roxb.) Hook. ex G.Don
Vanda testacea (Lindl.) Rchb.f.
Vanda vipianii Rchb.f.
Vandopsis gigantea (Lindl.) Pfitzer
Vandopsis shanica (Phillimore & Sm.) Garay

NEPAL / NÉPAL (LE) / NEPAL

Ascocentrum ampullaceum (Roxb.) Schltr.
Calanthe alismaefolia Lindl.
Calanthe alpina Hook.f.
Calanthe biloba Lindl.
Calanthe brevicornu Lindl.
Calanthe chloroleuca Lindl.
Calanthe densiflora Lindl.
Calanthe herbacea Lindl.
Calanthe mannii Hook.f.
Calanthe odora Griff.
Calanthe pachystalix Rchb.f.
Calanthe plantaginea Lindl.
Calanthe puberula Lindl.
Calanthe sylvatica (Thouars) Lindl.
Calanthe tricarinata Lindl.
Calanthe triplicata (Willem.) Ames
Rhynchostylis retusa (L.) Blume
Vanda alpina (Lindl.) Lindl.
Vanda coerulescens ssp. **boxallii** Rchb.f.
Vanda cristata Lindl.
Vanda pumila Hook.f.
Vanda stangeana Rchb.f.
Vanda tessellata (Roxb.) Hook. ex G.Don
Vanda testacea (Lindl.) Rchb.f.
Vandopsis undulata (Lindl.) J.J.Sm.

NEW CALEDONIA / NOUVELLE-CALEDONIE / NUEVA CALEDONIA

Calanthe balansae Finet
Calanthe hololeuca Rchb.f.
Calanthe × oreadum Rendle
Calanthe triplicata (Willem.) Ames
Calanthe ventilabrum Rchb.f.

NICARAGUA / NICARAGUA (LE) / NICRAGUA

Brassavola acaulis Lindl. & Paxton
Brassavola cucullata (L.) R.Br.
Brassavola grandiflora Lindl.
Brassavola nodosa (L.) Lindl.
Catasetum integerrimum Hook.
Catasetum maculatum Kunth
Miltonioides leucomelas (Rchb.f.) Bockemuhl & Senghas

NIGERIA / NIGÉRIA (LE) / NIGERIA

Aerangis biloba (Lindl.) Schltr.
Aerangis kotschyana (Rchb.f.) Schltr.
Angraecum angustipetalum Rendle
Angraecum angustum (Rolfe) Summerh.
Angraecum aporoides Summerh.
Angraecum birrimense Rolfe
Angraecum claessensii De Wild.
Angraecum distichum Lindl.
Angraecum egertonii Rendle
Angraecum eichlerianum Kraenzl.
Angraecum infundibulare Lindl.
Angraecum moandense De Wild.
Angraecum multinominatum Rendle
Angraecum podochiloides Schltr.
Angraecum pungens Schltr.
Angraecum pyriforme Summerh.
Angraecum subulatum Lindl.

PANAMA / PANAMA / PANAMÁ

Brassavola acaulis Lindl.
Brassavola nodosa (L.) Lindl.
Brassavola venosa Lindl.
Calanthe calanthoides (A.Rich. & Galeottii) Hamer & Garay
Catasetum bicolor Klotzsch
Catasetum maculatum Kunth
Catasetum planiceps Lindl.
Catasetum sanguineum Lindl. & Paxton
Catasetum viridiflavum Hook.
Miltonioides carinifera (Rchb.f.) Senghas & Luckel
Miltonioides schroederiana (O'Brien) Brieger & Luckel
Miltoniopsis roezlii (Rchb.f.) God.-Leb.
Miltoniopsis warscewiczii (Rchb.f.) Garay & Dunst.
Renanthera striata Rolfe
Rossioglossum powellii (Schltr.) Garay & G.C.Kenn.
Rossioglossum schlieperianum (Rchb.f.) Garay & G.C.Kenn.
Rossioglossum schlieperianum var. **flavidum** (Rchb.f.) Garay & G.C.Kenn.

PAPUA NEW GUINEA

Calanthe aceras Schltr.
Calanthe aruank P.Royen
Calanthe aurantimacula P. van Royen
Calanthe camptoceras Schltr.
Calanthe caulescens J.J.Sm.
Calanthe chrysoleuca Schltr.
Calanthe coiloglossa Schltr.
Calanthe cremeoviridis J.J.Wood
Calanthe cruciata Schltr.
Calanthe curvato-ascendens Gilli
Calanthe engleriana Kraenzl.
Calanthe finisterrae Schltr.
Calanthe fissa L.O.Williams
Calanthe hololeuca Rchb.f.
Calanthe inflata Schltr.
Calanthe kaniensis Schltr.
Calanthe labellicauda Gilli
Calanthe leucosceptrum Schltr.
Calanthe longifolia Schltr.
Calanthe lutiviridis P.Royen
Calanthe micrantha Schltr.
Calanthe parvilabris Schltr.
Calanthe pavairiensis P.Ormerod
Calanthe polyantha Gilli
Calanthe rhodochila Schltr.
Calanthe rhodochila var. **reconditiflora** (J.J.Sm.) S.Thomas
Calanthe stenophylla Schltr.
Calanthe sylvatica (Thouars) Lindl.
Calanthe torricellensis Schltr.
Calanthe triplicata (Willem.) Ames
Calanthe ventilabrum Rchb.f.
Calanthe vestita Lindl.
Renanthera edefeldtii F.Muell. & Kraenzl. ex Kraenzl.
Renanthera moluccana Blume
Renanthera philippinensis (Ames & Quisumb.) L. O.Williams
Vanda helvola Blume
Vanda hindsii Lindl.
Vandopsis lissochiloides (Gaudich.) Pfitzer
Vandopsis muelleri (Kraenzl.) Schltr.

PARAGUAY / PARAGUAY (LE) / PARAGUAY (EL)

Brassavola chacoensis Kraenzl.
Brassavola perrinii Lindl.
Miltonia flavescens (Lindl.) Lindl.

PERU / PÉROU (LE) / PERÚ

Brassavola chacoensis Kraenzl.

PERU (Continued)

?Brassavola nodosa (L.) Lindl.
Brassavola retusa Lindl.
Catasetum adremedium D.E.Benn. & Christenson
Catasetum barbatum (Lindl.) Lindl.
Catasetum callosum Lindl.
Catasetum coniforme C.Schweinf.
Catasetum cotylicheilum D.E.Benn. & Christenson
Catasetum fernandezii D.E.Benn. & Christenson
Catasetum hillsii D.E.Benn. & Christenson
Catasetum incurvum Klotzsch
Catasetum lanxiforme Senghas
Catasetum microglossum Rolfe
Catasetum monzonensis D.E.Benn. & Christenson
Catasetum moorei C.Schweinf.
Catasetum multifissum Senghas
Catasetum nanayanum Dodson & D.E.Benn.
Catasetum peruvianum Dodson & D.E.Benn.
Catasetum pleiodactylon D.E.Benn. & Christenson
Catasetum pulchrum N.E.Br.
Catasetum purusense D.E.Benn. & Christenson
Catasetum pusillum C.Schweinf.
Catasetum saccatum Lindl.
Catasetum schunkei Dodson & D.E.Benn.
Catasetum schweinfurthii D.E.Benn. & Christenson
Catasetum × sodiroi Schltr.
Catasetum tenebrosum Kraenzl.
Catasetum tenuiglossum Senghas
Catasetum transversicallosum D.E.Benn. & Christenson
Catasetum tuberculatum Dodson
Catasetum violascens Rchb.f. & Warz.
Miltoniopsis bismarckii Dodson & D.E.Benn.
Miltoniopsis phalaenopsis (Rchb.f.) Garay & Dunst.
Miltoniopsis roezlii (Rchb.f.) God.-Leb.
Miltoniopsis santanaei Garay & Dunst.
Miltoniopsis vexillaria (Rchb.f.) God.-Leb.

PHILIPPINES (THE) / PHILIPPINES (LES) / FILIPINAS

Ascocentrum aurantiacum (Schltr.) Schltr.
Ascocentrum aurantiacum ssp. **philippinense** Christenson
Calanthe angustifolia (Blume) Lindl.
Calanthe conspicua Lindl.
Calanthe davaensis Ames
Calanthe halconensis Ames
Calanthe hennisii Loher
Calanthe jusnerii Boxall ex Náves
Calanthe lacerata Ames
Calanthe lyroglossa Rchb.f.
Calanthe maquilingensis Ames

PHILIPPINES (Continued)

Calanthe mcgregorii Ames
Calanthe mindorensis Ames
Calanthe nivalis Boxall ex Náves
Calanthe pulchra (Blume) Lindl.
Calanthe rubens Ridl.
Calanthe triplicata (Willem.) Ames
Calanthe vestita Lindl.
Renanthera amabilis Boxall ex Náves (nomen)
Renanthera elongata (Blume) Lindl.
Renanthera matutina Lindl.
Renanthera monachica Ames
Renanthera philippinensis (Ames & Quisumb.) L. O.Williams
Renanthera storiei Rchb.f.
Rhynchostylis gigantea ssp. **violacea** (Lindl.) Christenson
Rhynchostylis retusa (L.) Blume
Vanda javierae D.Tiu ex Fessel & Luckel
Vanda lamellata Lindl.
Vanda limbata Blume
Vanda lindeni Rchb.f.
Vanda luzonica Loher ex Rolfe
Vanda merrillii Ames & Quisumb.
Vanda roeblingiana Rolfe
Vanda sanderiana Rchb.f.
Vandopsis lissochiloides (Gaudich.) Pfitzer

REUNION (FRENCH) / RÉUNION (FRANÇAIS) / REUNIÓN (FRANCESCA)

Angraecum borbonicum Bosser
Angraecum bracteosum Balf.f.& S.Moore
Angraecum cadetii Bosser
Angraecum calceolus Thouars
Angraecum caulescens Thouars
Angraecum cilaosianum (Cordem.) Schltr.
Angraecum cordemoyi Schltr.
Angraecum cornigerum Cordem.
Angraecum costatum Frapp. ex Cordem.
Angraecum crassifolium (Cordem.) Schltr.
Angraecum cucullatum Thouars
Angraecum divaricatum Frapp.
Angraecum eburneum Bory
Angraecum expansum Thouars
Angraecum expansum ssp. **inflatum** Cordem.
Angraecum filicornu Thouars
Angraecum germinyanum Hook.f.
Angraecum hermannii (Cordem.) Schltr.
Angraecum implicatum Thouars
Angraecum inapertum Thouars
Angraecum longinode Frapp. ex Cordem.
Angraecum macilentum Frapp.

REUNION (Continued)

Angraecum mauritianum (Poir.) Frapp.
Angraecum minutum Frapp. ex Cordem.
Angraecum multiflorum Thouars
Angraecum nanum Frapp. ex Cordem.
Angraecum oberonia Finet
Angraecum obversifolium Frapp.
Angraecum palmiforme Thouars
Angraecum parvulum Ayres ex Baker
Angraecum pectinatum Thouars
Angraecum pingue Frapp.
Angraecum pseudopetiolatum Frapp.
Angraecum ramosum Thouars
Angraecum salazianum (Cordem.) Schltr.
Angraecum spicatum (Cordem.) Schltr.
Angraecum striatum Thouars
Angraecum tenellum (Ridl.) Schltr.
Angraecum tenuifolium Frapp. ex Cordem.
Angraecum triquetrum Thouars
Angraecum undulatum (Cordem.) Schltr.
Angraecum viridiflorum Cordem.
Calanthe candida Bosser
Calanthe sylvatica (Thouars) Lindl.

RWANDA / RWANDA (LE) / RWANDA

Aerangis kotschyana (Rchb.f.) Schltr.
Angraecum infundibulare Lindl.
Angraecum marii Geerinck
Angraecum moandense De Wild.
Angraecum petterssonianum Geerinck
Angraecum sacciferum Lindl.
Calanthe sylvatica (Thouars) Lindl.

SAMOA / SAMOA (LE) / SAMOA

Calanthe alta Rchb.f.
Calanthe hololeuca Rchb.f.
Calanthe nephroglossa Schltr.
Calanthe triplicata (Willem.) Ames
Calanthe ventilabrum Rchb.f.

SAO TOME AND PRINCIPE / SAO TOMÉ-ET-PRINCIPE / SANTO TOMÉ Y PRÍNCIPE

Aerangis flexuosa (Ridl.) Schltr.
Angraecum aporoides Summerh.
Angraecum astroarche Ridl.
Angraecum doratophyllum Summerh.
Angraecum infundibulare Lindl.

SAO TOME AND PRINCIPE (Continued)

Calanthe sylvatica (Thouars) Lindl.

SENEGAL / SÉNÉGAL (LE) / SENEGAL (EL)

Aerangis biloba (Lindl.) Schltr.

SEYCHELLES / SEYCHELLES (LES) / SEYCHELLES

Angraecum calceolus Thouars
Angraecum eburneum Bory
Angraecum eburneum ssp. **superbum** Thouars
Angraecum multiflorum Thouars
Angraecum zeylanicum Lindl.

SIERRA LEONE / SIERRA LEONE (LA) / SIERRA LEONA

Aerangis biloba (Lindl.) Schltr.
Angraecum birrimense Rolfe
Angraecum distichum Lindl.
Angraecum multinominatum Rendle
Angraecum subulatum Lindl.
Calanthe sylvatica (Thouars) Lindl.

SINGAPORE / SINGAPOUR / SINGAPUR

Renanthera elongata (Blume) Lindl.
Rhynchostylis gigantea (Lindl.) Ridl.
Vanda insignis Blume

SOLOMON ISLANDS / ILES SALOMON (LES) / ISLAS SALOMÓN

Calanthe hololeuca Rchb.f.
Calanthe longifolia Schltr.
Calanthe pavairiensis P.Ormerod
Calanthe rhodochila Schltr.
Calanthe rhodochila var. **reconditiflora** (J.J.Sm.) S.Thomas
Calanthe triplicata (Willem.) Ames
Calanthe ventilabrum Rchb.f.
Renanthera edefeldtii F.Muell. & Kraenzl. ex Kraenzl.
Vanda hindsii Lindl.

SOUTH AFRICA / AFRIQUE DU SUD (L') / SUDÁFRICA

Aerangis mystacidii (Rchb.f.) Schltr.
Aerangis somalensis (Schltr.) Schltr.
Aerangis verdickii (De Wild.) Schltr.
Angraecum chamaeanthus Schltr.
Angraecum conchiferum Lindl.
Angraecum cultriforme Summerh.

SOUTH AFRICA (Continued)

Angraecum pusillum Lindl.
Angraecum sacciferum Lindl.
Angraecum stella-africae P.J.Cribb
Calanthe sylvatica (Thouars) Lindl.

SRI LANKA / SRI LANKA / SRI LANKA

Aerangis hologlottis (Schltr.) Schltr.
Angraecum zeylanicum Lindl.
Calanthe sylvatica (Thouars) Lindl.
Calanthe triplicata (Willem.) Ames
Rhynchostylis retusa (L.) Blume
Vanda spathulata (L.) Spreng.
Vanda tessellata (Roxb.) Hook. ex G.Don
Vanda testacea (Lindl.) Rchb.f.
Vanda thwaitesii Hook.f.

SUDAN (THE) / SOUDAN (LE) / SUDÁN (EL)

Aerangis kotschyana (Rchb.f.) Schltr.

SURINAME / SURINAME (LE) / SURINAME

Brassavola angustata Lindl.
Brassavola gardneri Cogn.
Brassavola martiana Lindl.
Catasetum × guianense G.A.Romero & Jenny
Catasetum longifolium Lindl.
Catasetum macrocarpum Rich. ex Kunth

SWAZILAND / SWAZILAND (LE) / SWAZILANDIA

Aerangis mystacidii (Rchb.f.) Schltr.
Angraecum pusillum Lindl.
Angraecum sacciferum Lindl.
Calanthe sylvatica (Thouars) Lindl.

TAHITI (FRENCH) / TAHITI (FRANCAIS) /TAHITI (FRANCÉS)

Calanthe tahitensis Nadeaud

TANZANIA (UNITED REPUBLIC OF) / RÉPUBLIQUE-UNIE DE TANZANIE (LA) / REPÚBLICA UNIDA DE TANZANÍA (LA)

Aerangis alcicornis (Rchb.f.) Garay
Aerangis brachycarpa (A.Rich.) Dur. & Schinz
Aerangis calantha (Schltr.) Schltr.
Aerangis carnea J.Stewart
Aerangis collum-cygni Summerh.

TANZANIA (UNITED REPUBLIC OF) (Continued)

Aerangis confusa J.Stewart
Aerangis coriacea Summerh.
Aerangis gravenreuthii (Kraenzl.) Schltr.
Aerangis hologlottis (Schltr.) Schltr.
Aerangis kirkii (Rchb.f.) Schltr.
Aerangis kotschyana (Rchb.f.) Schltr.
Aerangis luteo-alba (Kraenzl.) Schltr.
Aerangis luteo-alba var. **rhodosticta** (Kraenzl.) J.Stewart
Aerangis maireae la Croix & J.Stewart
Aerangis montana J.Stewart
Aerangis mystacidii (Rchb.f.) Schltr.
Aerangis oligantha Schltr.
Aerangis somalensis (Schltr.) Schltr.
Aerangis thomsonii (Rolfe) Schltr.
Aerangis verdickii (De Wild.) Schltr.
Angraecum brevicornu Summerh.
Angraecum chamaeanthus Schltr.
Angraecum conchiferum Lindl.
Angraecum cultriforme Summerh.
Angraecum decipiens Summerh.
Angraecum dives Rolfe
Angraecum eburneum ssp. **giryamae** (Rendle) Senghas & Cribb
Angraecum erectum Summerh.
Angraecum humile Summerh.
Angraecum minus Summerh.
Angraecum moandense De Wild.
Angraecum sacciferum Lindl.
Angraecum spectabile Summerh.
Angraecum stolzii Schltr.
Angraecum teres Summerh.
Angraecum viride Kraenzl.
Calanthe sylvatica (Thouars) Lindl.

THAILAND / THAÏLANDE (LA) / TAILANDIA

Ascocentrum ampullaceum (Roxb.) Schltr.
Ascocentrum curvifolium (Lindl.) Schltr.
Ascocentrum garayi Christenson
Ascocentrum pusillum Aver.
Ascocentrum semiteretifolium Seidenf.
Bletilla sinensis (Rolfe) Schltr.
Calanthe anthropophora Ridl.
Calanthe biloba Lindl.
Calanthe cardioglossa Schltr.
Calanthe ceciliae Rchb.f.
Calanthe clavata Lindl.
Calanthe densiflora Lindl.
Calanthe hirsuta Seidenf.
Calanthe labrosa (Rchb.f) Rchb.f.

THAILAND (Continued)

Calanthe lyroglossa Rchb.f.
Calanthe odora Griff.
Calanthe pulchra (Blume) Lindl.
Calanthe rosea (Lindl.) Benth.
Calanthe rubens Ridl.
Calanthe simplex Seidenf.
Calanthe succedanea Gagnep.
Calanthe sylvatica (Thouars) Lindl.
Calanthe tricarinata Lindl.
Calanthe triplicata (Willem.) Ames
Calanthe vestita Lindl.
Renanthera coccinea Lour.
Renanthera elongata (Blume) Lindl.
Renanthera isosepala Holttum
Renanthera matutina Lindl.
Renantherella histrionica (Rchb.f.) Ridl.
Rhynchostylis coelestis Rchb.f.
Rhynchostylis gigantea (Lindl.) Ridl.
Rhynchostylis retusa (L.) Blume
Vanda bensonii Bateman
Vanda brunnea Rchb.f.
Vanda coerulea Griff. ex Lindl.
Vanda coerulescens Griff.
Vanda denisoniana Benson & Rchb.f.
Vanda lilacina Teijsm. & Binn.
Vanda liouvillei Finet
Vanda pumila Hook.f.
Vandopsis gigantea (Lindl.) Pfitzer
Vandopsis lissochiloides (Gaudich.) Pfitzer

TOGO / TOGO (LE) / TOGO (EL)

Aerangis biloba (Lindl.) Schltr.
Angraecum multinominatum Rendle

TONGA / TONGA (LES) / TONGA

Calanthe hololeuca Rchb.f.

TRINIDAD AND TOBAGO / TRINITÉ-ET-TOBAGO (LA) / TRINIDAD Y TABAGO

Brassavola cucullata (L.) R.Br.
Brassavola nodosa (L.) Lindl.
Catasetum cernuum (Lindl.) Rchb.f.

UGANDA / OUGANDA (L') / UGANDA

Aerangis brachycarpa (A.Rich.) Dur. & Schinz

UGANDA (Continued)

Aerangis calantha (Schltr.) Schltr.
Aerangis collum-cygni Summerh.
Aerangis jacksonii J.Stewart
Aerangis kotschyana (Rchb.f.) Schltr.
Aerangis luteo-alba (Kraenzl.) Schltr.
Aerangis luteo-alba var. **rhodosticta** (Kraenzl.) J.Stewart
Aerangis thomsonii (Rolfe) Schltr.
Aerangis ugandensis Summerh.
Angraecum affine Schltr.
Angraecum distichum Lindl.
Angraecum erectum Summerh.
Angraecum firthii Summerh.
Angraecum infundibulare Lindl.
Angraecum moandense De Wild.
Angraecum reygaertii De Wild.
Angraecum sacciferum Lindl.
Calanthe sylvatica (Thouars) Lindl.

UNITED STATES OF AMERICA (THE) / ESTADOS UNIDOS DE AMÉRICA (LES) / ESTADOS UNIDOS DE AMÉRICA (LOS)

Brassavola cucullata (L.) R.Br.

VANUATU / VANUATU / VANUATU

Calanthe hololeuca Rchb.f.
Calanthe ventilabrum Rchb.f.

VENEZUELA / VENEZUELA (LE) / VENEZUELA

Brassavola angustata Lindl.
Brassavola cucullata (L.) R.Br.
Brassavola grandiflora Lindl.
Brassavola martiana Lindl.
Brassavola nodosa (L.) Lindl.
Brassavola retusa Lindl.
Catasetum barbatum (Lindl.) Lindl.
Catasetum bergoldianum Foldats
Catasetum bicallosum Cogn.
Catasetum bicolor Klotzsch
Catasetum callosum Lindl.
Catasetum collare Cogn.
Catasetum costatum Rchb.f.
Catasetum decipiens Rchb.f.
Catasetum discolor (Lindl.) Lindl.
Catasetum × **dunstervillei** G.A.Romero & Carnevali
Catasetum fimbriatum (Morren) Lindl. & Paxton
Catasetum gomezii G.A.Romero & Carnevali
Catasetum huebneri Schltr.

VENEZUELA (Continued)

Catasetum longifolium Lindl.
Catasetum macrocarpum Rich. ex Kunth
Catasetum maculatum Kunth
Catasetum maroaense G.A.Romero & C.Gomez
Catasetum merchae G.A.Romero
Catasetum microglossum Rolfe
Catasetum naso Lindl.
Catasetum parguazense G.A.Romero & Carnevali
Catasetum pileatum Rchb.f.
Catasetum planiceps Lindl.
Catasetum poriferum Lindl.
Catasetum rivularium Barb.Rodr.
Catasetum × **roseo-album** (Hook.) Lindl.
Catasetum sanguineum Lindl. & Paxton
Catasetum spinosum (Hook.) Lindl.
Catasetum × **tapiriceps** Rchb.f.
Catasetum × **wendlingeri** Foldats
Catasetum yavitaense G.A.Romero & C.Gomez
Miltonia spectabilis Lindl.
Miltoniopsis roezlii ssp. **alba** (W.Bull ex W.G.Sm.) Luckel
Miltoniopsis santanaei Garay & Dunst.
Miltoniopsis warscewiczii (Rchb.f.) Garay & Dunst.
Renanthera striata Rolfe

VIET NAM / VIET NAM (LE) / VIET NAM

Ascocentrum ampullaceum (Roxb.) Schltr.
Ascocentrum christensonianum Haager
Ascocentrum curvifolium (Lindl.) Schltr.
Ascocentrum garayi Christenson
Ascocentrum pusillum Aver.
Bletilla ochracea Schltr.
Calanthe alismaefolia Lindl.
Calanthe aleizettei Gagnep.
Calanthe angusta Lindl.
Calanthe angustifolia (Blume) Lindl.
Calanthe argenteostriata C.Z.Tang & S.J.Cheng
Calanthe brachychila Gagnep.
Calanthe cardioglossa Schltr.
Calanthe chevalieri Gagnep.
Calanthe clavata Lindl.
Calanthe densiflora Lindl.
Calanthe eberhardtii Gagnep.
Calanthe formosana Rolfe
Calanthe herbacea Lindl.
Calanthe lyroglossa Rchb.f.
Calanthe mannii Hook.f.
Calanthe odora Griff.
Calanthe petelotiana Gagnep.

VIET NAM (Continued)

Calanthe poilanei Gagnep.
Calanthe puberula Lindl.
Calanthe rubens Ridl.
Calanthe succedanea Gagnep.
Calanthe triplicata (Willem.) Ames
Calanthe velutina Ridl.
Calanthe vestita Lindl.
Renanthera annamensis Rolfe
Renanthera citrina Aver.
Renanthera coccinea Lour.
Renanthera imschootiana Rolfe
Rhynchostylis coelestis Rchb.f.
Rhynchostylis gigantea (Lindl.) Ridl.
Rhynchostylis retusa (L.) Blume
Vanda bidupensis Aver. & Christenson
Vanda concolor Blume
Vanda denisoniana Benson & Rchb.f.
Vanda lilacina Teijsm. & Binn.
Vanda lindeni Rchb.f.
Vanda liouvillei Finet
Vanda pumila Hook.f.
Vandopsis gigantea (Lindl.) Pfitzer

YEMEN / YÉMEN (LE) / YEMEN (EL)

Angraecum dives Rolfe

ZAMBIA / ZAMBIE (LA) / ZAMBIA

Aerangis appendiculata (De Wild.) Schltr.
Aerangis brachycarpa (A.Rich.) Dur. & Schinz
Aerangis collum-cygni Summerh.
Aerangis kotschyana (Rchb.f.) Schltr.
Aerangis montana J.Stewart
Aerangis mystacidii (Rchb.f.) Schltr.
Aerangis splendida J.Stewart & la Croix
Aerangis verdickii (De Wild.) Schltr.
Angraecum cultriforme Summerh.
Angraecum erectum Summerh.
Angraecum geniculatum G.Williamson
Angraecum minus Summerh.
Angraecum sacciferum Lindl.
Angraecum stolzii Schltr.
Calanthe sylvatica (Thouars) Lindl.

ZIMBABWE / ZIMBABWE (LE) / ZIMBABWE

Aerangis appendiculata (De Wild.) Schltr.
Aerangis kotschyana (Rchb.f.) Schltr.

ZIMBABWE (Continued)

Aerangis mystacidii (Rchb.f.) Schltr.
Aerangis somalensis (Schltr.) Schltr.
Aerangis verdickii (De Wild.) Schltr.
Aerangis verdickii ssp. **rusituensis** (Fibeck & Dare) la Croix & P.J.Cribb
Angraecum chamaeanthus Schltr.
Angraecum chimanimaniense G.Will.
Angraecum conchiferum Lindl.
Angraecum cultriforme Summerh.
Angraecum humile Summerh.
Angraecum minus Summerh.
Angraecum pusillum Lindl.
Angraecum sacciferum Lindl.
Angraecum stella-africae P.J.Cribb
Calanthe sylvatica (Thouars) Lindl.